Project Management for Facility Constructions

Alberto De Marco

Project Management for Facility Constructions

A Guide for Engineers and Architects

 Springer

Alberto De Marco
Politecnico di Torino
Department of Production Systems
and Business Economics
Corso Duca degli Abruzzi 24
10124 Torino
Italy
alberto.demarco@polito.it

ISBN 978-3-642-17091-1   e-ISBN 978-3-642-17092-8
DOI 10.1007/978-3-642-17092-8
Springer Heidelberg Dordrecht London New York

Library of Congress Control Number: 2011922051

Cover design: WMXDesign GmbH, Heidelberg

Printed on acid-free paper

Springer is part of Springer Science+Business Media (www.springer.com)

Contents

Chapter 1
Introduction

This handbook is a review of concepts, methods and practical techniques for managing projects to develop constructed facilities in the industries of oil and gas, power, infrastructure, architecture and commercial building.

It is addressed to a variety of professionals willing to improve their management skills. On the one hand, it is aimed at helping newcomers to the engineering and construction industry understand how to apply project management to the field practice. On the other, it allows experts in technical areas of engineering and construction approach project management disciplines.

In education, this text is suitable on undergraduate and graduate classes in architecture, engineering and construction management, as well as on specialist and professional courses in project management.

Project management is a professional practice involving a variety of disciplines to support the tasks required to effectively complete a project. Managerial activities include decision making, problem solving, planning, scheduling, directing, coordinating, monitoring and control.

In all sectors, projects are complex endeavors that call for the application of management practices from all players and stakeholders involved. In particular, plant and building asset construction projects require the joint effort of several actors usually organized on a multipart contract structure: owners, investors, lending institutions, developers, designers, construction contractors, and consultants, which take action with different perspectives and interest on the project.

The challenge is to establish a managerial environment that enables a successful project development while maximizing the mutual benefit of each party.

With this approach, I hope that this book will be a helpful brief guide for those who are asked to effectively contribute with various roles in a capital investment project.

1.1 Road Map

A project is a temporary enterprise distinguished for being complex and unique, with strict time, cost and quality objectives. In this notion, project management may apply to a large variety of projects in many production and service industries.

A. De Marco, *Project Management for Facility Constructions*,
DOI 10.1007/978-3-642-17092-8_1, © Springer-Verlag Berlin Heidelberg 2011

However, project management practices have to adapt to each specific context. One is the construction industry, referred to as the economic activities involving architecture, engineering and construction (AEC) of facilities, such as plant engineering (e.g. construction of pipelines, installation of power stations), real estate development, building construction (residential, industrial, commercial), and infrastructure (e.g. water treatment plants, roads and railways). Construction projects involve a period of time when physical deliverables are developed on a construction site: the implementation stage is the one requiring most of the resources over the project life-cycle.

For this reason, it is important that construction preliminaries such as organizing, design, and planning are performed to assure appropriate capacity, organization, and mechanisms for managing the basic resources of a construction project, namely: people, money and materials.

A project cannot be done without the combination of every one of these three components, and project management has to consider each limited resource pool as a definite area of management focus.

Human Resources management is about organizing a project-based company, creating knowledge and providing executive competences for effective project operations, establishing information exchange and communications between the people involved.

Money is referred to as project cash flow management. This includes the activities of evaluating and funding the project, budgeting cost and revenues along the project duration, measuring the actual expenditure, and controlling that the project is flowing according to the initial plans.

Material Resources management involves effective usage of construction materials, technologies and equipment, as well as the establishment of good processes for running the construction site.

These areas of management require two additional elements.

The first one is the *Contract* organization. People, money and material resources have to be properly organized within a contract framework. In fact, project management is inherent with running a set of contract agreements between different parties committed at various levels with financing, design, procurement and construction.

The second is *Uncertainty*. Very few projects are completed on time, in line with the expected budget and with the desired level of quality because of the risky and uncertain environment that human, financial and material resources have to face all along the project development stages.

This philosophy is illustrated in Fig. 1.1: Human, financial and material resources are areas of management to be organized within a contract framework and influenced by uncertainty.

This handbook is shaped accordingly and its scope is decomposed into five areas of management practice. Contents of each individual part are presented throughout the project life-cycle phases: feasibility, design and planning, construction monitoring and control, and close-out.

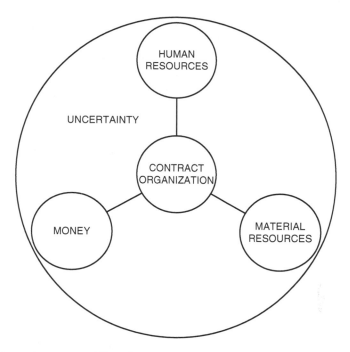

Fig. 1.1   Areas of management focus in context

Each part is introduced by a brief summary. Parts are divided into chapters, which include theory, examples, some practical case-studies, and a list of additional resources for further knowledge exploration.

1.2   Origin of the Book and Acknowledgments

This book collects contents from a variety of sources, books, papers, and contributions by various authors. The driving objective in writing this work was to collect from various original sources those components of knowledge that I consider very important for project managers and to present them as a brief hand guide for practitioners and senior students. Readers that want to have a deeper insight are recommended to explore the original quoted works.

Also, the text is mainly based on the course materials prepared for the 2006 Project Management class that I taught at the Engineering System Division, Dept. of Environmental and Civil Engineering at the Massachusetts Institute of Technology. The course was previously given by professors Fred Moavenzadeh, Nathaniel D. Osgood, and Keniosky Peña-Mora, whom I give grateful acknowledgement for originate course materials and resources. I also thank the students of the class for collaboration and contributions to the preparation of materials for this book, namely: Jose Paolo Calma, Cody L. Edwards, Patrick J. Hart, Masaki Ishii, Nikolaos S. Kontopoulos, David C. Lallemant, Marc Lopez, Andrew T. Lukemann, Joshua J. Maciejewski, Aaron S. Sarfati, Nicholas A. Shultz, Leticia Soto, Paul Sweeney, Matthew D. Williams, and Thaddeus Wozniak.

Above all, grateful acknowledge goes to professor Fred Moavenzadeh for continued support, interest in my work, reading and revising the book.

Sections of the book also build on materials developed for the Project Management class that I have been teaching at Politecnico di Torino (Italy) since 2007 as part of the Industrial Engineering and Management graduate program. I thank professor Carlo Rafele for opportunity, guidance and additional resources.

Part I
Contracting

The AEC industry is traditionally considered a captive market where projects are contracted based on spot customer-driven opportunities, long term planning is hard to enforce, and companies hardly fight to gain a competitive advantage through long range market positioning. In this context, whatever the role that a company plays in the industry, the management staff often perceives projects as one-off experiences with few linkages between project and corporate management.

On the contrary, program management and project portfolio management processes help to tighten this relationship and assist a company in the task of aligning single projects to the panel of other similar ones and, in turn, to achieve corporate strategic objectives.

Managing groups of projects in a coordinated way enables a better corporate financial control and strategic thinking with regard to several characteristics of a specific market, such as common approaches used to deal with customers, financial conditions, pricing, contracting systems.

Once the framework and practices of multiple project management are established in a company, individual projects are managed through effective contract organization and administration.

This is because project management is consistent with the execution of a contract as an obligation of the parties to respect agreed objectives of time, cost and quality. Also, this is because to meet the original goals of a construction agreement is the best assurance that the project cash flow will align with the one planned by the program managers and portfolio manager.

With this philosophy in mind, this part of the book includes:

– The basic principles to consider in managing a program of similar projects and a project portfolio (Chap. 2),
– The notion of contract organization and the most used construction contracting mechanisms (Chap. 3),
– The most important issues with regard to the process of administrating a construction contract (Chap. 4).

Chapter 2
Multiple-Project Management

2.1 The Multiple-Project Environment

A complex project-based business demands strong managerial effort to keep various types of overlapping projects continuously aligned to the changeable strategic goals of the organization.

To this end, a portfolio of projects and one or more programs have to be carried out. The portfolio is the ensemble of all programs the company is committed to.

A program is defined as a group of related projects managed in a coordinated way to obtain benefits not available from managing them individually (Project Management Institute 2008a). In the usual corporate language, a program is a medium/long term initiative that encloses two or more similar projects.

More properly, projects must be managed as interrelated efforts at different levels within the organization: a single project is assigned to a Project Management team; a group of similar projects is directed by a Program Management staff; and a collection of programs, the Project Portfolio, is managed by the corporate top-executive level (Fig. 2.1).

Usually all three of the tasks are supported by a Project Management Office (PMO) or an equivalent central staff.

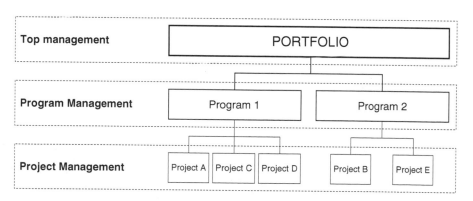

Fig. 2.1 Management levels of responsibility with projects, programs and portfolio

A. De Marco, *Project Management for Facility Constructions*,
DOI 10.1007/978-3-642-17092-8_2, © Springer-Verlag Berlin Heidelberg 2011

2.2 Project Portfolio Management

Planning a Project Portfolio requires construction organizations to define the corporate strategy and defining the strategic goals of the company. The strategy is usually, but not exhaustively, concerned with ameliorating the market positioning, creating new options for business expansion, and finding ways to optimize the business cost for enhancing value to the shareholders and the system of stakeholders (interested communities, customers, employees, suppliers). This is influenced by both external factors, such as market demand or competitors, and internal ones, like availability of human and material resources and various constraints.

The role of project portfolio management is define and control programs of similar projects with the purpose of reaching predetermined strategic targets.

In particular project portfolio managers are demanded to:

- categorize programs;
- evaluate the value and the risk that each program brings to the organization;
- compare, select and prioritize programs

2.3 Program Management

2.3.1 Notion of Program Management

"Program Management" is a definition that can be used with various meanings depending on the context.

Some companies appoint a Program Manager to take responsibility over project time scheduling and monitoring. Typically, this person works as a member of the Project Management Office (PMO) or as part of a large project team.

In other property management oriented situations, program management is referred to as the project management process carried over the total life-cycle of a facility, rather than during the design and construction phases only. The Program Manager is concerned with various activities from feasibility studies to occupancy, including operations and maintenance, if applicable. In this notion, the Program Manager is in staff to the owner or is an independent professional acting on his behalf with the task of integrating the various project management roles that are demanded all along the facility life-cycle, namely: designer, construction manager, contractor's project manager, and operations manager.

Finally, the emerging definition of Program Management is consistent with the notion of multiple-project environment. While a Project Manager is assigned the tasks of planning, controlling and directing a single project, the Program Manager keeps an eye on several projects at once, acting as a planner, controller and director of a group of two, or more, similar projects.

The first two definitions are, at some extents, part of all Project Management functions and competences, as presented in the following chapters. In these sections, we place emphasis on the third definition of Program Management, referred to as management of a folder of similar or related projects

2.3.2 Grouping Projects

Depending on complexity, size, and role of the organization, which may act as owner, contractor or professional service provider, the multiple-project environment poses problems that result from different market orientation, the varied nature of projects, limited resources and competition between projects (Archibald 2003).

Thus, programs can be defined according to market destination, average size of projects, product type, geographical area, or delivery system.

For example, a general contractor may decide to group projects by type of client industry. In this case, programs are managed as business units: e.g. power, oil and gas, chemical and pharmaceuticals, transportation, environment, industrial plants, etc.

Otherwise, it may want to consider all small-medium projects as a single program, while large-scaled or complex projects as part of a separate program.

Splitting programs by product type is also very common among contractors; they manage a building program separately from a road program, even if buildings include a mix of industrial, commercial, or residential construction.

Worldwide or global organizations often group projects depending on their location, because of the different framework of local culture, regulations and practices (e.g.: North-America, Latin America, Western Europe, Middle East, etc.).

Finally, similarities in projects can be found in the type of delivery system with no difference between products: all projects delivered as turnkey contracts may be part of a unique family needing the same standards, processes and resources. If a reliable contractor is strategically chosen by an owner to perform turnkey constructions, the sequence of projects' development may depend on the limitations the contractor has on available resources that are skilled for the program.

While executives are responsible for grouping projects within appropriate value-oriented programs aimed at responding to strategic directions, program managers are concerned with selecting project opportunities and with providing best practices and guidance for managing the projects included in the program.

Archibald (2003) states that the higher-order objective of Program Management is to reach the overall strategic goals of the organization by supporting project managers in the process of reaching the successful completion of their projects.

Program managers have to bridge the gap between corporate strategy and project execution. To this end, their action has to provide project managers with the people and toolbox to carry out similar projects within a program.

This may include:

- a PMO, also called Program Office, to support proposal management, planning, scheduling, time/cost monitoring and reporting by providing competent project managers and staff, and making methods and information tools available for the task;
- a centralized procurement and allocation service of human resources, capital and construction materials;

- an integrated knowledge management system;
- communication interfaces between projects, programs and top management.

2.3.3 Defining Programs: Selecting and Prioritizing Projects

It often happens in business that projects are initiated due to operating necessity, because the organization needs to run or simply because the boss wants them. Sometimes projects are selected for competitive necessity, to face capacity expansion and not to lose market shares, or because they are opportunity-driven. But most often, the selection of investment opportunities arises from the need of giving the highest return to limited financial resources.

To this end, construction program managers have to select and rank similar projects by using objective assessment methodologies. The selection process can be based on qualitative or quantitative methods, or a combination of both.

Qualitative techniques may be used to compare advantages and disadvantages. The SWOT analysis, which lists strengths, weaknesses, opportunities and threads on a table, is typically used to broadly understand the project challenges and to make a first screening between similar projects in order to pick the ones with maximum advantages, and discard the ones that are likely to bring the highest prospect risks.

Qualitative aspects of projects may also be compared with multiple attribute techniques. The method is based on the notion of expected utility: individuals make choices to maximize their implicit preference (Haimes 2009; see also Chap. 13). Multiple attribute scoring techniques enable decision-makers to give each option a preference index and to compare alternatives (Karydas and Gifun 2006).

To proceed with this semi-qualitative method, the Program Team or PMO defines a set of attributes that are important for the decision. Each attribute is assigned a utility weight (by way of a scale, e.g. from 0 to 100%) as a degree of priority of the specific attribute for making the decision. Then, projects are scored with respect to every one of the attributes with a numerical value that is an indicator, in the decision maker point of view, of the project capability of satisfying the specific attribute. Finally, the performance index of a given project (k) is the summation, for all parameters (i), of the utility weight (w) multiplied by the score (s) of each attribute, as in Eq. (2.1) (Karydas and Gifun 2006).

$$PI_k = \sum_i^n w_i * s_{ik} \qquad (2.1)$$

To compare projects, qualitative methods may not be enough and it becomes necessary to introduce numerical evaluation of the return on investment by measuring their profit and profitability.

Profit is referred to as the gross profit of a project (the total revenues deducted the direct cost), and return equals the gross margin over revenue or over cost.

Profitability evaluations are based on discounted project cash flow, using net present value (NPV) and internal rate of return (IRR) computations. See Chap. 8 for details about the determination of discounted indices and ratios.

Thus, multiple attribute methods involving both quantitative and qualitative measures can be more properly used to select projects where impacts cannot be easily and purely estimated in terms of dollar amounts. Frequently decision-makers care about multiple attributes: namely cost, time, quality, relationship with owner, impact on health, safety and environment, public and commercial image, etc.

Alternatively, a semi-qualitative Performance Index (PI) and a numerical index of profitability, such as the NPV, can be kept separate by making a tradeoff decision. Even if we cannot directly weigh one attribute against another one to sort out an objective priority, we can rank some consequences about the examined similar projects. At least, we can rule out projects giving consequences that are inferior with respect to all attributes: it is defined that these projects are "dominated by" others. The notion is strictly derived from the decision making theory defining a decision as "Pareto optimal" (or efficient solution) if it is not dominated by any other decision.

The key concept here is that we may not be able to identify the best projects, but we can discard bad ones. The in Fig. 2.2 example is drawn to better show the

Project	PI	NPV
A	5.0	0
B	5.5	0.2
C	6.5	0.2
D	7.0	0.3
E	6.5	0.4
F	8.0	0.1
	9.0	0.5

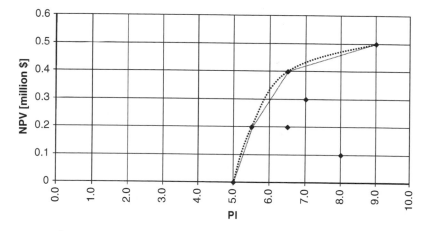

Fig. 2.2 Project selection using a multiple-attribute tradeoff decision making approach

selection process in a double tradeoff project selection, considering both qualitative performance index and the NPV of seven similar projects.

In the example, it is obviously clear that projects C, D, and F are non efficient investments, because they are dominated by project E, which maximizes the NPV while maintaining the same performance index of qualitative attributes. More widely, all projects in the bottom-right side of the chart (inside the "efficient frontier" represented by the curve) are wasteful investments compared to the others. This technique can be readily extended to additional dimensions.

Quantitative methods for comparing and valuing projects also include benefit-cost and cost-effectiveness analyses, commonly used for public projects. In such cases cost discounting still often applies.

Benefit-cost analysis tries to consider both the economic and non-economic benefits, such as social or environmental impacts over time. It takes into account all accrued costs, also those resulting by assuming that the project was not built. A benefit/cost ratio ranks projects as acceptable if it is greater than one (i.e. benefits greater than costs). Problems of objectivity arise because the benefit/cost ratio often fails to consider the absolute size of the benefits, and it can be difficult to determine whether something counts as a benefit or a negative cost (MIT Open Courseware).

Cost-effectiveness is a fairly similar method, but avoids assigning a monetary value to all non-economic factors, such as "lives saved", or "quality of life". Instead, the summary of a project is specified in terms of ratios of gross margin per non-monetary quantity ("deaths averted", "quality-adjusted life years saved"). These ratios provide a means of comparing the relative non-monetary benefits provided from a limited pool of money. This notion is gaining value in some municipalities and governmental bodies for prioritizing infrastructural investments and as a consideration in regulatory design.

2.3.4 Developing Programs

Once the projects are appropriately selected, a program must be developed and effectively managed.

A simplistic framework suggests that a successful program management depends on the following activities:

- scheduling projects according to their interrelations, if applicable. This approach requires the usage of network-based scheduling techniques that consider the highest levels of breakdown structures of different projects;
- estimating and forecasting resource requirements and usage according to the overlapping of projects;
- continuous monitoring and re-scheduling to take into account risks and changing conditions.

In general, the program management process is developed according to the following main steps.

- The program master plan gives strategic directions, categorizes and selects those projects that fulfill the strategic outlook, defines objectives and the general scope of the project portfolio. It may be compared to the planning phase of project management. In this step the program management team provides guidelines to execute each project: work breakdown structure, schedule, scope, quality and communication standards.
- The program master schedule settles the time constraints for each project. It is usually provided as a large Gantt chart showing start and finish dates of each project and including the main milestones of the program.
- The program budget collects all information about the cost of projects. By matching the schedule and the budget, it is possible to obtain the expected program cash flow and, therefore, to define the financing resources.
- Once the program has been planned and scheduled, it can be executed. So, the program management team is given the role to support the project management for each initiative and in each knowledge area of management focus: integration, scope, time, cost, quality, human resources, communications and procurement management (Project Management Institute 2008b).
- During execution of the projects it is necessary to manage the program control process as a whole: the overall program performance results from each project performance with regard to time, cost and quality. The communication and reporting activities about the entire program are easier if the information from single projects is centralized and available in real-time.

References and Additional Resources

Archibald RD (2003) Managing high-technology programs and projects, 3rd edn. Wiley, Hoboken, NJ

Bennet J (1991) International construction project management: general theory and practice. Heinemann, Butterworth

Haimes YY (2009) Risk modeling, assessment, and management, 3rd edn. Wiley, Hoboken, NJ

Karydas DM, Gifun JF (2006) A method for the efficient prioritization of infrastructure renewal projects. Reliab Eng Syst Saf 91:84–99

Levy SM (2000) Project management in construction. McGraw-Hill, New York, NY

Massachusetts Institute of Technology, Open Courseware, Project Management Class. mit.edu/ocw/

Project Management Institute (2008a) A guide to the project management body of knowledge, 4th edn. Project Management Institute, Newtown Square, PA

Project Management Institute (2008b) Standard for project portfolio, 4th edn. Project Management Institute, Newtown Square, PA

Springer ML (2001) Program management: a comprehensive overview of the discipline. Purdue University Press, West Lafayette, IN

Chapter 3
Contract Organization

3.1 Roles in Construction Projects

In principles, a construction project is the outcome of a joint effort between the owner, one or more contractors, and various professional entities that provide finance, design, construction, and operation services. These entities work as parties within a contract framework with mutual relationships.

At a glance, owners may be subsumed into three broad areas.

- *Service owners* include public and private entities that build an infrastructure, a facility or a utility project with the purpose of running the business with a long-term social or economic return on investment. Examples of this kind of owner are: a municipality to build a new school or a road, an electric utility company to erect a new power station, and a hotel company to run its own accommodation facilities.
- *Property* or *asset managers* act as landlords that develop building investments to get long-term return from rental fees and facility management services. An example is an insurance company or a private-equity real estate investor.
- *Real estate developers* aim at selling constructed facilities to the market with a short-term return on investment. Typically, real estate developers have residential housing and office building programs.

Real estate developers and property managers do business within the construction industry, while service owners are usually industrial players in a specific consumer or business-to-business market.

Owners usually go into an agreement with contractors that have specialization to construct the product required by the project, such as building, infrastructure and civil works, or plant engineering. In the previous examples, the municipality will look for a building contractor; the electric utility company will select a specialist engineering and contracting firm; the real estate developer will join a residential housing constructor.

Contractors also differentiate if they have design capabilities. Most general contractors are responsible for the construction job based on design specifications and drawings produced by owner's architects and engineers. Yet, some contractors are

A. De Marco, *Project Management for Facility Constructions,*
DOI 10.1007/978-3-642-17092-8_3, © Springer-Verlag Berlin Heidelberg 2011

able to supply both design and construction, thus acting as design-build firms, or engineering-procurement-construction (EPC) contractors.

Many contracting firms have production capacity to develop part of the construction with own human resources and site equipment, while portions of the job are subcontracted to smaller specialist trade contractors. Usually from 20 to 80% of the job is assigned to specialized trade subs. In addition, contractors with a solid financial capacity are likely to participate in the funding effort both on short-term and long-term projects through equity financing.

From a life-cycle point of view, contractors may offer construction operations and maintenance (O & M) services. Such long-term contractors usually have also facility O & M capabilities.

Finally, construction projects request a large variety of professional entities and service providers, such as finance, legal, design, and construction assistance.

- *Financial and legal services* consist of project financing, bonding, and contract administration. Such services are provided by lending institutes, insurance companies, lawyers, and consultants.
- *Design* includes a large variety of architecture and engineering services from feasibility to post-construction. Typical services are basic design, process engineering, detailed engineering, construction permits, code and regulation compliancy, shop drawings, and as-built drawings.
- *Construction assistance* is a term to identify various professional activities with regard to project and construction management, safety inspection, quality assurance, site supervision and direction. As better presented later into the book, construction project management services involve planning, scheduling, direction, coordination, procurement, monitoring and control of the project.

3.2 Notion of Contracting

The definition of a proper construction contracting system, aimed at fitting the characteristics and goals of a project, is the first and main tool to correctly allocate risks, responsibilities, duties and rights between the parties involved. The choice of such a contracting system has major ramifications throughout the life time of the project.

For the purpose of discussing about many possible kinds of construction contract arrangements, in the literature contracts are described as composed of three main components: a delivery system together with a payment scheme and an award method.

The delivery system defines the nature and number of the portions of project scope – design, construction, and finance – that are contracted to each business entity and the organizational relationships between the parties. The payment scheme defines how the owner will pay the contractor. The award method, aimed at selecting the contracting parties, establishes the rules for assigning the contract.

An owner must go through a rational decision process to combine delivery system, payment type and award method into the desired and appropriate contracting organization for his project. Choosing a contracting system is not a precise task. In some cases there is no one single best method, but several that may successfully apply with advantages and disadvantages (Gordon 1994, pp. 197–198).

3.3 Delivery Systems

The scope of work of a capital project can be roughly segmented into financing, design, and construction.

As far as financing is concerned, usually a short-term loan covers the initial investment required for developing the project, and a subsequent long-term financing is reimbursed with operations profits (additional information is in Chap. 7).

Design is a process involving further detailing of scope from broad project concept to technical specifications and ready for construction (RFC) drawings. The task is usually sequenced into a basic design followed by a detailed engineering stage. The output of basic design is used to secure construction permits and authorizations, while detailed engineering is precursor to construction execution.

Here, construction includes pre-construction activities (such as site preparation) and physical implementation.

A contractor may be responsible for part of one, two or all three of the scope components.

Typically, either separate organizations perform each portion of the job, or a sole entity develops integrated design and construction either as a single constructor with design capabilities or as a joint venture between an engineering firm and one or more constructors.

In principles, there are four possible main scenarios, as shown in Fig. 3.1.

- Separate design and build phases: the contractor is only responsible for the physical construction of the specified project by own or subcontracted labor, materials procurement and site management tasks. Design is conducted by the owner, directly or through a delegated architecture/engineering firm (A/E), and financing is arranged by the owner. This involves a traditional Design-Bid-Build (DBB) sequenced process: the owner awards the construction contract based on detailed engineering documents prepared by the A/E. Then a general contractor or more prime contractors are managed directly by the owner's project management team or an early-hired Construction Manager to support in pre-construction and construction development.
- Integrated design-build: the contractor takes responsibility for both design and construction of the facility. The contract is typically awarded based on approved basic design, but sometimes bidders have to submit a proposed basic design developed according to the prescriptions of a feasibility study that specifies the owner's needs and performance requirements. The financing is made available by the owner.
- Integrated design, build and finance: in addition to design and physical development, the contractor is also asked to contribute to either short-term funding,

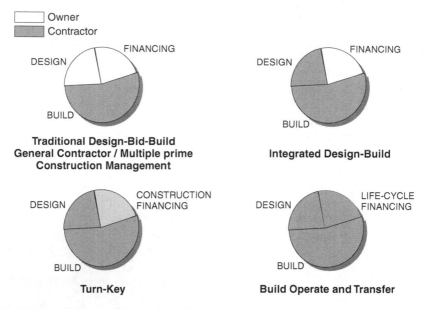

Fig. 3.1 How different delivery systems allocate the scope of work to the contractor

as in turnkey agreements, or life-cycle financing as in build-operate-transfer arrangements (Gordon 1991).

Following is a description of the main construction delivery systems.

3.3.1 Design-Bid-Build

Design Bid Build contracts (DBB) are largely used and very familiar to most western construction industries.

In these cases, the owner engages the professional services of an architect/engineer to develop all stages of design and to control construction performed by a general contractor (GC), which in turn may subcontract part of the scope of contract to selected specialist traders (Fig. 3.2).

Subcontracting practice has largely increased since the "1990s, because of augmented sophistication of construction technology, globalization of large companies looking for local construction capabilities, cheaper cost and reduced risk the GC has to bear.

The sequential DBB process requires a high level of collaboration between the project participants, which have different marginal interests in the contract:

- the owner is concerned with budget respect, timely completion, quality satisfaction and site safety,

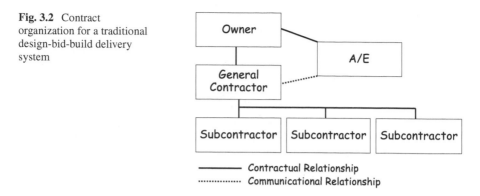

Fig. 3.2 Contract organization for a traditional design-bid-build delivery system

- the A/E is mostly interested in the quality, aesthetics and technical performance of the constructed facility, recognition of his work, and in limiting his liabilities;
- the general contractor is typically most interested in leveraging revenue and cutting quality to assure maximum profit, within acceptable limits related to maintaining his market reputation.

Diverging perspectives often result in adversarial relationships, which demand rigorous application of project management practices and collaboration between all players.

The traditional contracting method has several advantages. First, it is a well known mechanism that has long been used by owners, contractors, designers, and courts. Also, it is of practical value if project uncertainty exists primarily in design: no construction is developed before a slow-paced and flexible design process is completed.

However, the method has many disadvantages with regard to the following issues.

- Constructability. Design constructability is traditionally not thoroughly reviewed before actual construction, resulting in design and construction changes, in minimization of contractor's knowledge and capabilities, and in the loss of opportunities for time and cost savings.
- Fast-tracking schedule. With DBB, there is no way to expedite the project by overlapping design and construction.
- Changes and budget. After the contract has been awarded, any changes to the original contract detailed design impose heavy additional cost for the owner, who is also highly dependent on the contractor for the quality of the job (and this is why often the owner employs on-call contractors to complete the scope of work). Thus, the budget is far from a fixed price, and often rises as construction activities unfold: general contractors often seek changes to make extra profit, which in turn leads to extra time and cost for resolving disputes.

3.3.2 Construction Manager

As a peculiarity of the US and UK markets, the Construction Manager[1] (CM), in its pure or "Agency" form, is a business entity acting as a project manager and as a construction consultant to the owner.

A Construction Manager professional staff is early hired:

- to support the designer with planning and pre-construction tasks, such as constructability review of design, value engineering, estimation, alternative selection, schedule, financing, management of the design team, and early procurement of long lead-time items;
- to break down the scope of work into a number of elements contracted to specialist trades;
- to manage the competitive selection processes of specialist contractors on behalf of the owner;
- to accelerate the project through the use of fast-tracking, in which construction is commenced before design is complete;
- to be a common reference point and to act as a facilitator in conflicts between owner and contractors;
- to provide quality assurance, coordinate work of sub/contractors, manage change orders and claims, perform inspections, and assure safety conditions on the construction site;
- to care of project monitoring and control, job and management meetings, reporting, as well as various administrative tasks.

If the CM service is executed directly by the owner with no use of a construction consultant, the delivery mechanism is called "Multiple Primes". The general contractor is eliminated and replaced with a manager (also called Owner's Project Manager) responsible for selecting and managing more than one construction contractors.

The organization chart in Fig. 3.3 shows the relationships in a Pure or "Agency" CM mechanism.

CM started in the late 1960s in response to the increased need of an experienced organization to better control chronic cost and schedule overruns.

The system has some general disadvantages, in the sense that there may be potential conflicts between CM, designers and contractors, and the fast-track process does not allow to define the final cost when construction starts with little confidence that the project will be completed on budget.

In addition, the use of fast-tracking elevates risks of discovery of design specification inconsistencies during construction, and the need for rework and design changes at that time.

---

[1] The Construction Manager is often termed as a "Management Contractor" in the UK.

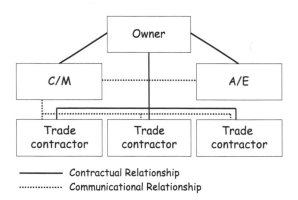

Fig. 3.3 Contract organization for an agency CM delivery system

This mechanism also puts little incentive on the CM to reduce price and completion time, because the owner alone takes the risk of the cost of the project without guarantees from the CM.

To overcome these drawbacks, today specific CM firms are willing to shoulder the risk of cost overrun out of indefinite estimates of the total final cost of a project. These firms act as "Construction Manager at risk". With the purpose of reducing the owner's financial exposure, the "CM at risk" entity is given responsibility of all contractual relationships with subcontractors, thus transforming the CM into a sort of general contractor hired since the design phases. Most times, an agreed-upon-the-contract Guaranteed Maximum Price (GMP) gives the owner assurance that the project will not overcome the budget because all cost overruns out of the ceiling GMP will be paid solely by the CM.

The Guaranteed Maximum Price (GMP) is typically agreed when design is 60–90% complete, to give the CM sufficient details to produce a reasonable and fair cost estimate and GMP.

Both a high fee and strict performance bonds are introduced to mutually satisfy both parties.

With the risk approach, CM has direct contract relationship with trade contractors and is sole responsible of selecting, managing and paying them (Fig. 3.4).

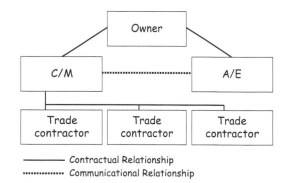

Fig. 3.4 Contract organization for a CM at risk delivery system

However, the GMP is often hard to enforce, as it sets a defined price for an undefined final product of the contract. So, the CM is no longer impartial and can make claims during the construction phase for changes that are out of the original guaranteed scope.

Also, hiring the CM at risk early during design may result in design pressure, reduced quality of final outcome and more price risk, while a late CM contract may not maximize constructability and production capabilities. Thus, the CM at Risk, acting as a contractor, may originate tensions and adversarial relationships with the owner.

3.3.3 Design-Build

The Design-Build delivery system is broadly used for private projects in most western countries with little differences. For public works, the process is not still expressly allowed in some European countries, as well as in some US states.

Under this mechanism (Fig. 3.5), the owner develops feasibility studies and design concepts to define needs and functional requirements, and then makes the deal with a sole entity with both design and construction skills.

The company that will undertake the project can be either a Design-Build (BD) firm or a joint venture formed for this specific purpose, such as a consortium. The DB firm may also subcontract the design work.

This system, known in the plant construction industry as Engineering Procurement and Construction (EPC) contract, works very well for complex and sophisticated projects, but requires phased design to protect all parties from extreme risks.

As in the CM case, tension for when to recruit DB firm is motivated by a tradeoff: if the DB is early hired, it is hard to objectively select it, while a later recruit gives less benefit, because it reduces fast-track ability and design creativity. A rigorous selection process has to take into consideration design, price, and schedule.

DB has several advantages: it allows fast tracking, a single point for accountability and coordination, and handing of complex technology.

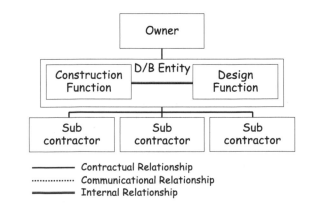

Fig. 3.5 Contract organization for a design-build delivery system

However, the contract price is difficult to formulate and enforce before design is developed. Moreover, the owner lacks a fiduciary relationship with the A/E, which may in turn lead to low-quality design to shield the contractor's profit.

To avoid this, the method demands a sophisticated owner who will stay on top of the design process in order to supervise and ensure to get the requested quality. In very complex projects, the method can end up without the desired result if the owner is not closely involved.

Also, the owner, once the project is underway, cannot get rid of or pick up individual team members (e.g. special subcontractor that would influence the task in a positive or negative manner). In the DB system there is also little space for over-checks and balances, which means that many problems may be hidden until late (no A/E or CM supervising), or that the final product may not satisfy the initial needs. Finally, if fast track is used, it can result in much redesign, reworks, iterations and completion delays.

A variation of the DB procedure that is being used in some countries is known as "Bridging". It is a process where two A/E entities work, with one under the owner's call and the other as a service to the contractor. The owner appointee defines the functional and aesthetic characteristics of the project (basic design), while leaving leeway for contractors to do detailed design. The contractor's A/E does the final construction drawings under the supervision of the owner's A/E. Typically, construction cannot begin until the final RFC drawings are completed. The lack of fast track, while giving the appearance of requiring additional time, assures no misunderstandings about what was intended by the initial drawings of the owner's designer.

Despite all the cons, the bridging system is a relatively new practice within the US construction market and it will take some time to be considered as a widely acceptable technique. Most often, owners in that market prefer to appoint a Construction Manager with the task of supervising the DB contractor, rather than using a bridge A/E (Fig. 3.6).

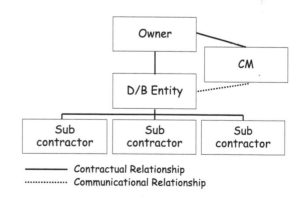

Fig. 3.6 Construction manager supervising the design-build contractor

——— Contractual Relationship
············· Communicational Relationship

3.3.4 Turnkey

In a Turnkey process the contractor executes the whole design and build work and also short-term finances the project during the construction period. A lump-sum payment is due at the time of commissioning the final constructed facility to the contractor. Often, a small advance payment or a few milestone payments may be negotiated to be reimbursed whenever major portions of scope of work are substantially completed and checked out for quality compliance.

The method is very common in residential housing and plant construction (e.g. power and oil plants). This method gives the owner the time to raise funds while construction is underway.

3.3.5 Build-Operate-Transfer

One of the major objectives of a contract is to enable economic satisfaction for all parties involved, or, at least, balancing economic expectations on a reasonable fair compromise. Yet, litigation and conflicts are typical in construction contracts, where owners and constructors seek economic return with detriment to the other party's outcome and, in turn, to the project.

One way to overcome the problem is to establish strategic collaborative relationships between owner, contractors and service providers to enable win-win conditions. Typically, collaboration works very well under either financial risk-sharing or long-term contract agreements (Bennet 1991), such as in the Build-Operate-Transfer (BOT) delivery system, where long-term financing covering the operations and maintenance (O & M) period is defined.

With this type of agreement, the contractor is responsible for financing the project through a special purpose vehicle company (SPV). The client owns the property, but provides the contractor with exploitation rights on long term operations (usually from 20 to 60 years). After that period, the owner gets back the facility for its own O & M and usage, typically with no extra cost.

End users pay directly the operating SPV for facility usage, such as in the use of a toll road highway, toll bridge, or an energy utility.

If the constructed facility does not give a straight and fair compensation to the SPV, or is overly risky, the owner can contribute to the initial investment or pay an annual fee to assure profitability to the concessionaire company. In this latter context, the delivery system is more properly called Build-Lease-Transfer: the owner cannot afford the initial investment, but is willing to substitute it with a long lease.

There are many different aspects of this system but most of them are used for Public-Private Partnerships (PPP) to construct public infrastructures and buildings all around the world. Famous PPP large-sized projects are the rail tunnel under the Channel and the EuroDisney amusement park.

3.3.6 Summary of Delivery Systems

As a conclusive note, different delivery systems have to cope with distributing the project risk between owner and contractor. The way the parties are capable and eager to take over risk defines the construction delivery system.

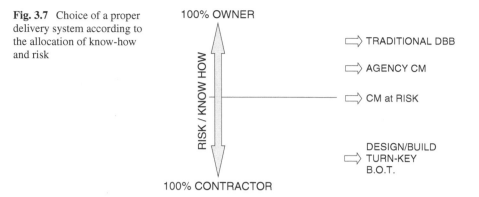

Fig. 3.7 Choice of a proper delivery system according to the allocation of know-how and risk

Figure 3.7 gives a scheme on who shoulders the risk in the various possible options.

The more the project know-how is held by the owner, the more he is willing to take risks and act directly as the project manager (Owner's Project Manager) or appoint a fiduciary manager (Agency Construction Manager). For instance, in traditional DBB the owner has the project managed by the designer.

Accordingly, the more the contractor is aware of the crucial role she is playing, the more financial risks she will be able to manage by using his own construction management organization and capabilities over the project development period (CM at risk, D/B, Turnkey) or over a long time frame involving O & M capacities (BOT). This practice goes under the name of Project Management by contractor.

3.4 Payment Schemes

As anticipated, a key idea for creating cost savings lies in establishing an appropriate contractual framework for risk sharing between owner and contractor.

Different parties differ in their ability to manage or tolerate various types of risk: an owner (or a big contractor) often better handles geotechnical risks or weather risk; contractors often better manage risk of slow teams, equipment quality, procurement, and quality of supervision.

A successful agreement divides risks to save money on contract price and provides incentive to contractors to have them finish early, in budget, and with good quality. Such incentives are strongly influenced by the mechanism used by the owner to pay the contractor for work performed. Thus, a correct payment scheme has to go with the choice of the appropriate delivery system to enable the participants' commitment in cost savings.

Even though the notion of risk is the focus of Sect. 3.5 coping with uncertainty in project management, here it is important to highlight how an agreement takes "risk premiums" into account to define a proper contract organization.

Contractors are often highly risk averse; the contractor is willing to "pay" the owner a risk premium (i.e. charge less for contract) if the owner assumes certain risks:

- for risks that contractor can't control, it may be willing to pay a risk premium to the owner to assume such risks. The contractor here will lower the price if the owner takes on such risks (essentially, paying the owner a risk premium);
- for risks that contractor can control, it will be cheaper for the contractor to manage them than to pay a risk premium to the owner.

The fundamental direction for saving cost is to structure the contract so that risks better handled by the contractor are imposed on contractor, and risks the owner can easily handle are kept by the owner.

The fundamental balance is to impose:

- high enough risk incentive to get the contractor do his job efficiently, within the contract provisions (e.g. incentive to finish on time, incentive to stay within budget);
- impose low enough risk on the contractor to have reasonably low bid;
- impose risk according to the contractor's capacity to tolerate risk.

The concept of risk premiums has derivative implications for accountability and monitoring.

Let us consider parties A and B in an agreement: the greater the risk to party A, the higher the incentive on it to manage this risk and the lower on party B to handle it. This provides incentive on A to monitor the relevant factors so that A can act to promptly manage any risk that is materializing and so that B can't claim the risk is responsible for a problem.

Finally, the greater the risk assumed by A, the greater the incentive on B to make sure that A's means of risk management fall within the agreement (e.g. that A is not "cutting corners", over reporting material quantities required, or otherwise cheating to shield itself from risk).

Construction price and timing are also affected by risk premium. The greater the level of uncertainty and risk imposed on contractor, either the longer will the contractor be tempted to delay construction until uncertainties play out, or the larger the amount the contractor will charge up front.

However, the owner usually seeks to minimize the up-front cost. Thus, the solution here is dual. On the one hand, it is possible to lower uncertainty by further design stages and by having the owner shoulder the risk of possible changes.

On the other hand, the owner can expedite the works by paying a higher price to the contractor as a premium for taking on pressure and risks of changes and timely construction.

Fig. 3.8 Payment schemes depend on the risk allocation between owner and contractor

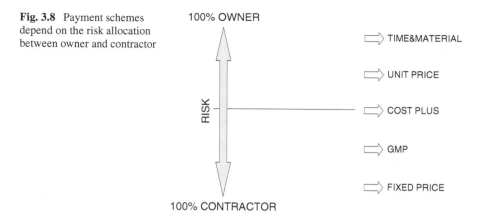

Risk allocation is a crucial issue. Depending on the business intent, the property owner might want to carry all of the risk in order to save on price, or pay risk premiums to have contractors shoulder the project risks.

Figure 3.8 depicts various possible payment schemes depending on risk allocation between the parties.

When the owner uses a time and material or a unit price payment scheme, he shoulders most of the financial risk associated with the project.

Negotiated cost-plus-fee contracts may have different provisions regarding compensation of the contractor, usually based on risk sharing policies between the contract parties.

When the owner uses a firm-fixed-price or guaranteed-maximum-price payment scheme, the contractor bears most of the financial risk.

We briefly review each of these payment schemes below.

3.4.1 Time and Material

In a time and material contract, the contractor is reimbursed for all actual expenses for direct cost (labor, material, equipment), and paid a percent fee that includes overhead cost and a fair profit:

$$\text{Contract price} = (\text{labor} + \text{material} + \text{equipment}) * (1 + \% \text{ fee}) \qquad (3.1)$$

Because labor prices are usually defined according to unionized wages including standard overhead and profit, or are agreed upon the contract, the project price equals:

$$\text{Contract price} = \text{labor} + (\text{material} + \text{equipment}) * (1 + \% \text{ fee}) \qquad (3.2)$$

Of course, all costs are based on detailed worksheets and bill of materials, as shown in the following example.

Scheduled working time:
  Specialized worker   75.00$/h; 50 h
  Worker   50.00$/h; 100 h

Expected usage of material:
  Equipment rental   5,000$
  Construction commodities   3,000$
  Contractor's overhead   30%

Contract value $= (75.00*50 + 50.00*100) + (5,000 + 3,000)*1.3 = 19,150\$$

As a result, the risk is put on owner: the actual contract value is defined only at completion, the owner's budget is not guaranteed and the contractor has no incentive to proceed quickly and effectively since a longer execution project duration leads to increased revenues. Typically a high contingency buffer is considered by the owner to face possible cost overruns out of the original project estimates.

The time and material payment scheme works very well for small urgent projects with a high level of uncertainty (e.g. emergency repair and maintenance works), which inevitably leads to ongoing adjustments.

3.4.2 Unit Prices

Under the common Unit Pricing scheme, the contractor agrees to be paid the unit price of each specified item of work. Each unit price usually includes direct cost, as well as overhead cost and profit. Sometimes overhead items, such as construction site equipment, are paid separately.

If the project has a number of items and activities, it needs a detailed list of unit prices; if the project is homogeneous or linear (i.e. with very few items, such as a tunnel excavation), it can be described by only one unit price which includes the amount of different activities (e.g.: unit price per meter of completed tunnel, unit price per cubic meter of poured concrete).

In the case of a DBB, the estimated contract quantities are listed by the A/E as part of the request for proposal documents, and the unit prices are those quoted by the contractor into the bid.

However, the total sum of money due to contractor for each item is unknown until construction is finished, because payment is made based of measured work performed. Therefore, the unit price contract requires the owner to measure the actual quantities by keeping on site the owner project representative or a consultant A/E acting as a quantity surveyor.

Quantity influences price because of economies of scale for procurement or work rate, so that typically the unit price is renegotiated if quantity deviation is 10–20% off, according to the contract clauses.

The following example includes two items in the scope of work: procurement and erection of pre-casted concrete footings and columns.

Items:
    Footings   80.00$/sq ft
    Columns   1,550.00$/unit

Scheduled quantities:
    Footings   100 sq ft
    Columns   9 units

Contract initial value $= 80 * 100 + 1,550 * 9 = 14,750\$$

Unit pricing is a valuable payment scheme to get a low bid, and it requires the owner only to keep track of performed quantities. Yet, this highly depends on the accuracy of the estimation of contract quantities, otherwise leading to cost overruns. In fact, the total cost for the owner can be greater than planned.

On the other side, the contractor can make profit because payment is based on actual quantities, but he can also lose money in the same way.

Also, unbalancing of bids may cause additional expense to the owner from disproportionate cash flow. Unbalancing a bid means that if contractor believes actual quantity of a particular item will differ, he increases and/or decreases the unit price in anticipation of that deviation. Also, a bid may be intentionally unbalanced to get an earlier payment from the owner: this can be done by overpricing early items and underpricing later ones, thus covering early project costs and contractor's cash out. Consequently, the owner has to keep an eye on unbalanced bids and exclude a contractor if its bid is highly unbalanced.

Unit pricing is an interesting example of risk sharing: the owner takes risk for uncertainty in quantity, while contractor bears the risk for increased cost of individual items, as a result of differing actual efficiency of work rate or procurement cost compared to bid estimates.

3.4.3 Cost Plus Fixed Percentage Fee

Similarly to the time and material payment scheme, with a Cost plus a fixed percentage-of-cost fee the owner reimburses the contractor for all direct and project overhead actual expenses and pays a percent profit on top of cost. Differently from time and material, here the reimbursable cost includes all billings, such as for services and subcontracts.

The contractor agrees to execute the contract scope of work, while shouldering very little risk. Indeed, he has little commitment to cost saving because the greater the cost, the greater the absolute profit he will be getting from the project, because of the fixed percentage, as shown in Fig. 3.9.

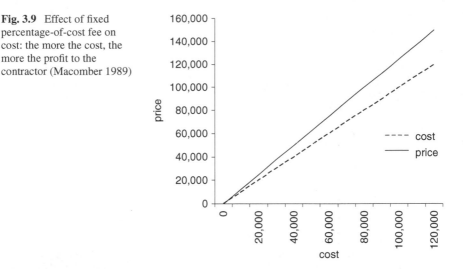

Fig. 3.9 Effect of fixed percentage-of-cost fee on cost: the more the cost, the more the profit to the contractor (Macomber 1989)

The following example also shows how the contractor has no incentive to reduce cost, because he will not only enjoy the same percent return on cost (ROC) from both reduced (scenario A) and increased cost (scenario B), but also be earning more money if cost is higher than planned.

Initial contract value	
Estimated cost	$100,000
Fixed % fee	8%
Contractor's Return on Cost ROC	8%

Actual contract value at completion:
- *Scenario A: Cost saving*
  Actual cost $90,000
  Contractor Price = 90,000 *(1+8%) = $97,200
  R.O.C. = 8%
- *Scenario B: Cost overrun*
  Actual cost $110,000
  Contractor Price = 110,000 *(1+8%) = $118,800
  R.O.C. = 8%

The advantages of cost plus fixed percentage fee are the following.

- It assures maximum flexibility to the owner with no disputes over change orders because the contractor gets paid for any extra work requested by the owner.
- It permits collaboration at early project stages, minimal negotiation time, and minimal fear of commitment by contractor.
- The owner only has to pay for what the project actually costs. If he closely manages the project, he can save money.

Disadvantages are the following:

- Contractors have incentive to grow the scope and price of contract.
- The owner shoulders all risk.
- There is little incentive to the contractor to reduce costs, and overtime salaries can even increase costs.
- Costs are unknown until the contract is closed.
- The lack of risk on the contractor forces the owner to shoulder the effort in risk monitoring, and may lead to a low-quality project. The owner needs to oversee construction closely, speed up slow crews, and identify management problems.

3.4.4 Cost Plus Incentive Fee

To overcome these disadvantages the owner needs to incentivize the contractor to reduce time and cost. This in turn requires that the contractor is given ways to handle these issues.

Incentives the contractor can benefit may be related to different aspects of project performance, such as schedule, quality, cost, safety, and other factors. For example, bonus-penalty arrangements can be used with regard to time of contract completion, such as a bonus paid for each day of early completion and a liquidated damage charged for each day of late completion.

A proper and common way to introduce incentives is to negotiate a refined payment scheme: the main types of such incentives include cost-plus-fixed-fee, target-cost-plus-incentive-fee, and guaranteed maximum price payment schemes (American Management Association 1986; Gordon 1991). These schemes are discussed below.

3.4.5 Cost Plus Fixed Fee

In cost-plus-fixed-fee arrangements the contractor is paid the actual cost plus a fee as a fixed amount of money. The actual cost of the project may be different than the budgeted one, but the fee remains firm. Design should be completed or sufficiently advanced to define a reliable estimate of the project cost and a fair fixed fee on top.

This payment scheme drives early finishing, because a longer duration of construction increases indirect cost and reduces profitability. Thus, this type of payment scheme is opportune to provide incentives to the contractor when time of completion is of great importance to the owner. Yet, it is recommended that there are no penalties on delays, to avoid cutting corners and litigation between the parties.

Meanwhile, the contractor also bears risk for growing size of project: the greater the cost, the less the relative return this will get from the project, as shown in Fig. 3.10.

The following example introduces a fixed fee to the same scenarios that were analyzed in the case of a cost-plus-percentage-fee payment scheme.

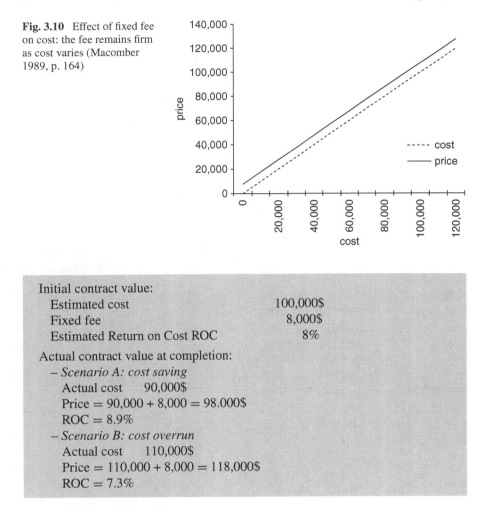

Fig. 3.10 Effect of fixed fee on cost: the fee remains firm as cost varies (Macomber 1989, p. 164)

Initial contract value:
  Estimated cost   100,000$
  Fixed fee   8,000$
  Estimated Return on Cost ROC   8%

Actual contract value at completion:
  – *Scenario A: cost saving*
    Actual cost   90,000$
    Price = 90,000 + 8,000 = 98.000$
    ROC = 8.9%
  – *Scenario B: cost overrun*
    Actual cost   110,000$
    Price = 110,000 + 8,000 = 118,000$
    ROC = 7.3%

3.4.6 Target Cost Plus Incentive Fixed Fee

When the budget cost is of great importance to the owner, the cost-plus-fixed-fee contract can be arranged to have the contractor make his best effort to maintain the actual cost within a certain target.

The incentive is to have the contractor share either the extra cost or the savings out of the target, or both.

If the contractor makes savings on the total target cost, he has right to share the saved money (usually from 20 to 50% of savings). If he spends more than the target cost, he shares an agreed fraction of cost overrun. The actual contract price at completion will equal:

$$\text{Price} = \text{AC} + (\Delta\text{C} * \% \text{ share}) + \text{firm fixed fee} \qquad (3.3)$$

where AC is the actual cost, and ΔC equals target minus actual cost.

The following example includes both provisions: the contractor is always paid the actual cost plus or minus a share of savings or additional costs.

Contract provisions:
    Target cost   100,000$
    Fixed fee   8,000$
    Savings or extra-cost share   30%
    Estimated Return on Cost   ROC = 8%

Actual contract value at completion:
    – *Scenario A: cost under target*
    Actual cost   90,000$ (cost saving = +10,000$)
    Price = 90,000 + 10.000 * 0.30 + 8,000 = 101,000$
    ROC = 12.2%
    – *Scenario B: cost over target*
    Actual cost   110,000$ (cost overrun –10,000$)
    Price = 110,000 – 10.000 * 0.30 + 8,000 = 115,000$
    ROC = 4.5%

By assuming that the price equals the actual cost, which means to set the formula (3.3) equal to AC, it is possible to determine the maximum AC that gives profit to the contractor; for a greater actual cost, the contract will provide a loss.

For the given example, the maximum actual cost allowing profit is:

$$\text{Price} = AC + (100,000 - AC) * 0.30 + 8,000 = AC \rightarrow AC = 126.666.70\$$$

Figure 3.11 shows the profit depending on value of actual cost.

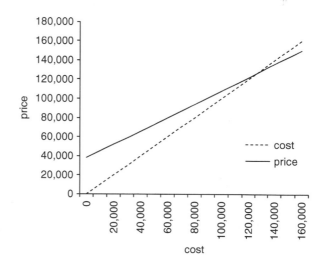

Fig. 3.11 Effect of variation of actual cost from 0 to 160,000$ to profit

3.4.7 Cost Plus an Award Fee

Another form of incentive is a cost-plus-an-award-fee payment scheme mainly used when there are uncertain conditions. A base fee is established as a percentage of the target cost and guaranteed to the contractor for completion of the contract. To the base fee an additional award fee is paid depending on quality, time and cost performance.

This payment scheme is of great advantage: it motivates contractors both because they are paid the work performed and a percentage profit (thus allowing flexibility to accommodate uncertainty in scope) and because an extra profit is rewarded for good work.

3.4.8 Guaranteed Maximum Price (GMP)

GMP is a variation of cost-plus-fixed-fee: the contractor is reimbursed the actual cost of work performed plus a fixed fee, up to a prearranged ceiling on price. Out of the GMP, the contractor assumes any additional costs, and the owner is assured his budget will not be exceeded.

A shared-savings contract provision may apply to GMP contracts, usually between 30 and 60%, as shown in the following example and Fig. 3.12.

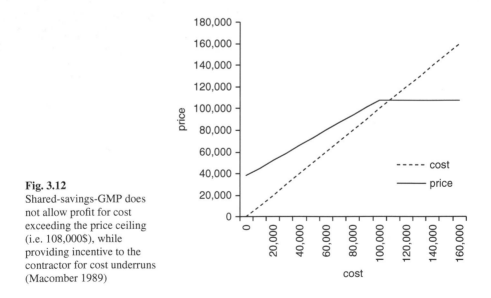

Fig. 3.12
Shared-savings-GMP does not allow profit for cost exceeding the price ceiling (i.e. 108,000$), while providing incentive to the contractor for cost underruns (Macomber 1989)

Contract provisions:
  Estimated cost   100,000$
  Fixed fee   8,000$
  GMP   108,000$
  Saving share   30%
  Estimated Return on Cost   ROC = 8%

Actual contract value at completion:
  – *Scenario A ($10,000 in savings) = cost + shared savings + fixed fee*
    Actual cost   90,000$
    Price = 90,000 + 10.000 * 0.30 + 8,000 = 101,000$
    ROC = 11.000/90.000 = 12.2%
  – *Scenario B ($5,000 in extra costs) = cost + fixed fee*
    Actual cost   105,000$
    Price = 105,000 + 8,000 = 113,000$ > GMP
    so that Price = GMP = 108,000$
    ROC = 3.000/105.000 = 2.8%
  – *Scenario C (extra costs over ceiling) = GMP*
    Actual cost   110,000$
    Price = 110,000 + 8,000 = 118,000$ > GMP
    so that Price = GMP = 108,000$
    ROC = –2.000/110.000 = –1.8% (loss)

Because the contractor takes most of the financial risk, the GMP scheme works relatively well when the contractor holds the project know-how and directly defines the scope of contract, such as in design-build or turnkey delivery systems.

The advantages of GMP are as follows:

- it permits easier financing for the owner because the maximum price is known since the beginning;
- the owner keeps either part or the total of savings below the GMP;
- it allows fast-tracking, particularly if the contractor is already involved with design.

Disadvantages are as follows:

- the price may be high if design is not complete, as in most delivery systems;
- the contractors may still not tightly control costs;
- the owner must monitor contractor spending.

3.4.9 Firm Fixed Price

Fixed price payment scheme dictates that an overall fixed amount of money is paid to the contractor for the total scope of contract. Here, all materials, labor, equipment, overhead cost and profit are comprised into the price with little possibility for the contractor to claim for cost and schedule overruns, unless the contract provides for reimbursable scope changes initiated by the owner.

This payment system is clearly helpful to owners because they know the cost of the project before it begins.

However, it minimizes risk for the owner only if the project is well estimated and the contract documents clearly accurate. Otherwise, disputes may arise over changes that may bring the price far above the initially anticipated contract value.

The fixed-price scheme is a high incentive to contractors to finish early (so that they can move on to other jobs) and at low cost because the lower the cost, the greater the profit, as shown in the Fig. 3.13.

However, high is the risk that the contractor will seek schedule and cost savings with detriment to project quality.

A fixed-price is usually lower than a GMP because savings are not shared by the owner.

Fixed-price payment is sometimes also called lump-sum payment. Yet, a lump-sum is referred to as a stipulated amount of money that covers all aspects of the work described by the contract documents, however it has been determined. Under this wider definition, a contract lump-sum may be computed on a time and material, unit prices, or fixed-price basis.

The common aspect of such lump-sum contracts is that the owner selects the contractor based on a competitive bidding: the scope of work is well defined to the level of detailed design so that bids are accurately priced as a comparable total sum.

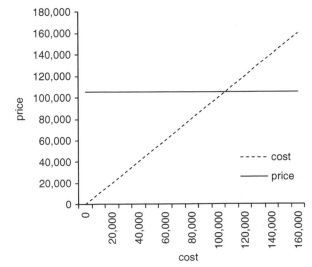

Fig. 3.13 Fixed-price as cost varies

The different type of payment schemes will only determine the reliability of the bid lump-sum: with time and material, as well as with unit prices, the value of the contract is most likely to change during the project execution for changes in the scope of work, while in fixed-price it will likely remain firm. This explains why fixed-price can also be used when design is not complete and the owner needs to bid the contract.

3.4.10 Summary of Payment Schemes

As a summary (Fig. 3.14), it is useful to compare how the different payment schemes give incentive to contractors to complete the job efficiently and at low cost, from time and material, unit prices and cost-plus-fixed-percentage-fee, where contractors increase profit as a result of increasing costs; through cost-plus-fixed-fee, where contractors have incentive to expedite construction in order not to reduce profit; to GMP and fixed-price, where contractors achieve profit by reducing cost and project duration.

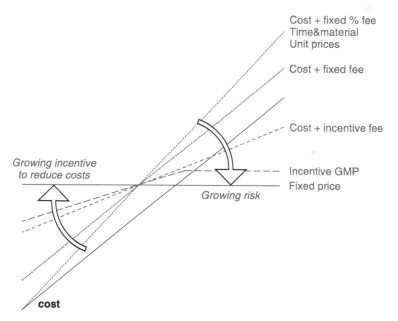

Fig. 3.14 Incentives (relative to the contractor) compared in different payment schemes. The centre point is the sum of cost plus a fee

The risk is shared accordingly: the more the incentive to make profit on savings, the more the contractor will shoulder the risks of additional costs, as a result of the capability of a given payment scheme to manage changes. By the same token, elevated risks to a contractor on a project will typically lead to higher contractor bids or reservation prices in negotiation.

3.5 Award Methods

The method that can be used to award the contract is of crucial importance to make the delivery system and the payment scheme work effectively together.

The contractor may be selected based either on competition or negotiation.

Owners make use of open competitive bidding to get to the lowest lump-sum price and market transparency. Yet, access may be restricted to bondable and recorded bidders. Award may go to low bidder, but most often contractors are ranked based on an arithmetic combination of multiple factors, such as:

- qualification and price;
- schedule and price;
- qualification, schedule and price proposal;
- design and price, for those delivery systems where design is part of the contractor's scope of contract.

While it allows for taking advantage of a good price and a transparent process, competitive bidding can set up a win-lose situation between the owner and the contractor: competitive pressures, such as time pressure resulting in an insufficient consideration of design before pricing and downwards cost pressure due to fear of being underbid by the competition and project risks impose the danger of narrow- or non-existent contractor profit margins. Unfortunately, any cost savings secured by the owner that result from such competitive pressures can be dwarfed by the costs – both tangible and intangible – associated with elevated rates of change orders, cutting corners and dispute-oriented relationships.

Different tradeoffs are involved with regard to the time provided to submit proposals and the number of bidders. If the time provided to bidders to review the contract documents is too short, there is a high risk of low-quality bids, which incorporate a high risk premium or are unrealistically low, and the potential that too few bidders will be willing or able to participate. In spite of the short-term delay of construction that is required, providing bidders more time to review documents may reduce the schedule overall, due to reduced problems in the course of the project.

A different tradeoff relates to the number of bidders involved. Moreover, if there are too many bidders, the process is likely to scare away the best contractors; while if bidders are too few, the process is insufficiently competitive.

In government bids, the process is usually based on standard local regulations; while private biddings have lots of variations. Generally, the typical public bidding process is managed according to the following instructions:

- it is overseen by an A/E firm;
- it is advertised in newspapers and other form of publications, where qualification requirements are specified;
- bid documents are provided after advertising and usually include design, a fair cost estimate, and sample contract;

- qualification occurs after submission of bids in public bidding; while a private bidding is usually by invitation only, and qualification occurs before submission of bids;
- before proposals are submitted, owners have to answer to "requests for information (RFI)" and/or hold a pre-bid conference to explain the scope of contract, working conditions, and answer questions
- typically, a 60-day period is provided to prepare and submit bids.

In a negotiation, contract provisions, risk allocation, payment scheme, price and schedule are negotiated between the owner and one or more qualified contractors. Typically, negotiation is used either for very simple projects, when a trusted and familiar party is involved, or for very complex projects, when there is the need of a special experienced contractor, involved in design, and starting the work early. Negotiation is usually associated with design-build and turnkey delivery systems using cost plus incentive, GMP, and fixed-price payment schemes.

The aim of the negotiation is to find the balance point of the parties' mutual satisfaction (Thompson 2004). Generally, negotiation requires a savvy owner and contractor able to get a win-win agreement because of differences in risk preferences in price and other attributes, within compliant constraints. A win-win agreement is defined as the most acceptable one for both parties.

With additional attributes, negotiations become more complex. Briefly, the key skill in multiple-issue negotiations (such as those considering more than one attribute, i.e. price, schedule, quality, design, etc.) is to find a "Pareto optimal" point within the range of different constraints and expected utilities of owner and contractor. In other words, the ranges represent the risk premiums that owner and contractors have to share or impose the other party for taking on the specific risk.

Of course, there is more than one Pareto optimal agreement, so that multiple-issue analysis of possible agreements represents a means to exclude the dominated solutions and to define possible incentives and contract clauses that represent mutual gains.

3.6 Selecting the Appropriate Contract Organization

Several are also the methods proposed by both academic (Gordon 1994; Hendrickson 2008; Clough and Sears 1994) and professional institutes to define the suitable contract architecture that may apply to a specific capital project.

Advice in the topic is provided by the Association of General Contractors of America (Dorsey 1997), the American Institute of Architects and the American Society of Civil Engineers. These methods provide description of the possible contract organizations and many drivers to consider for defining the appropriate ones. Drivers include flexibility of design, time constraints, design interaction with construction and financial approaches. The Construction Industry Institute (Oyetunji and Anderson 2002) also gives practical tools to develop a structured procedure and decision support tool to aid owners in selecting a project delivery and contract strategy.

As a general approach to the problem, the combination of a delivery system together with a payment scheme should be based on a risk allocation policy and on the prioritization of quality, time and cost constraints for a specific project (De Marco and Rafele 2008).

Delivery systems, payment schemes and award methods may be combined in a variety of contract arrangements. Yet, based on the principles of allocating the risk on the party that is more likely to easy handle it, it is suggested that a delivery system is associated with the corresponding payment schemes.

For example, a traditional DBB system may be appropriately combined with a time and material or unit price contract, while a design-build system is more suitable for incorporating a fixed price. The agency CM may be paid based on cost plus fixed fee to give him small risk, while still discouraging from increasing cost as a result of currently used percentage fee on top of cost. A CM at risk is appropriate if this goes with a cost-plus-incentive-fee with a GMP.

Similarly, the award method has to follow the risk/know-how metrics as above: in a traditional DBB system with unit prices, there is the opportunity of looking for the low bid; while in D-B contracts, it is more opportune that incentive payment schemes are negotiated in details.

With the purpose of providing a simple method and assuring an aligned contract strategy for the parties involved, it is proposed that the varied criteria for selecting a delivery system and payment mechanism are reduced to the main goals of a project, namely: time, cost and quality.

Time is consistent with the objective of a speed construction process and the respect of a contract baseline. In the owner perspective this gives surety that the constructed facility will be available for occupancy and operations according to the business plan. For the contractor, a short duration of the project may require accelerated extra cost to perform timely construction with harm to the profit. As a consequence, the contractor will be interested in pursuing a fast process only under appropriate economic incentives. Cost is inherent with budget respect. The owner is typically dedicated to containing cost within acceptable limits while contractors may pursue extra cost with resulting increased profit if they are not participating to the owner's benefits of a reduced cost. Quality is related to the satisfaction of contract technical specifications and requirements, but also with flexibility in design and construction changes. It is very hard that all three objectives are reached together in a project execution. This usually involves a tradeoff between them. For example, project acceleration and quality improvements usually require additional resources and cost overruns.

It is required that the owner defines priority face to each of the goals, and makes decision about the appropriate combination according to the precedence of factors. Following is a summary chart that can be used for an aligned contract mechanism (Fig. 3.15), where a larger grey area indicates the ability of the contract arrangement to fulfill the given objectives.

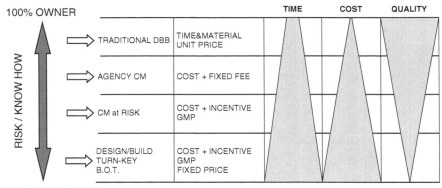

Fig. 3.15 Possible contract arrangements based on the main project objectives (De Marco and Rafele 2008)

Figure 3.15 shows that a traditional DBB with reimbursable pricing policy is suitable only if quality is assigned maximum priority: the owner may make changes and get the best quality while the contractor will likely accept improvements since those are appropriately reimbursed as an incentive to quality commitment. It is stated that this system will be practicable only under the strict management from the owner. If the budget respect is the prior goal, then owner's direction may conflict with the capacity of the contractor to expedite the process with his own management. Under a D/B contract paid with incentive-fee the contractor will underperform quality to give cost and time the appropriate precedence. The CM system may balance all attributes by appropriately performing professional project management services.

As a conclusive note, aligning both the owner's and the contractor's interests to the aims of a project is of great importance to the project's success. This can be done by appropriately selecting a contract arrangement as a combination of a delivery system, a payment scheme and an award method. The selection process has to take into account the risk allocation between the owner and the contractor as a result of the know-how each party has in designing and constructing the project. Also, since it is not possible to get the maximum quality with minimum time and low cost, the three project goals have to be assigned a priority. The contract mechanism will take into account both risk sharing and priority of project objectives as drivers for a contract strategy enabling alignment of the project participants.

All over the world, associations and independent organizations help project participants in this task and define contract standards for both local and international construction contracts.

The International Federation of Consultant Engineers (FIDIC) defines worldwide recognized contract guidelines for international agreements, such as the New Engineering Contracts (NEC). In the US markets, the Associated General Contractors of America (e.g. AGC 400 series), the American Institute of Architects

(AIA), the Construction Management Association of America (CMAA), as well as professional engineering organizations provide guidelines for procurement of construction services, including forms of contractual provisions, arrangements, obligations and responsibilities of parties. These standard contracts are recommended for use as a guideline for the formulation of a specific contract, and not to be directly sourced as a template.

References and Additional Resources About Contract Organization

American Management Association (1986) The contracts management deskbook, revised edn. New York, NY
Anderson S, Oyetunji A (2003) Selection procedure for project delivery and contract strategy. Construction Research 1–9
Bennet J (1991) International construction project management: general theory and practice. Heinemann, Butterworth, pp 351–353 and 356–377
Clough RH, Sears GA (1994) Cocvnstruction contracting, 6th edn. Wiley, New York, NY
De Marco A, Rafele C (2008) Aligning construction project participants on appropriate contract arrangement. Proceedings of 22nd IPMA World Congress, Rome, 9–11 November 2008, vol 1, pp 195–200
Dorsey RW (1997) Project delivery systems for building construction. Associated General Contractors of America, New York, NY
Fisk ER (2003) Construction project administration, 7th edn. Pearson Education, Upper Saddle River, NJ
Gordon CM (1991) Compatibility of construction contracting methods with projects and owners. MIT MS thesis, http://hdl.handle.net/1721.1/12955
Gordon CM (1994) Choosing appropriate construction contracting method. J Construc Engin 120(1):196–210
Hendrickson C (2008) Project management for construction, 2nd edn. Carnegie Mellon University, Pittsburgh, PA, http://www.ce.cmu.edu/pmbook/
Kavanagh TC, Muller F, O'Brien JJ (1978) Construction management a professional approach. McGraw-Hill, New York, NY
Macomber JD (1989) You can manage construction risk. Harv Bus Rev 67(2):155–161
Nicolas JM, Steyn H (2008) Project management for business, engineering, and technology, 3rd edn. Elsevier, Burlington, MA
Oyetunji A, Anderson S (2002) Project delivery and contract strategy. Research Report 165-12. Construction Industry Institute, Austin, TX
Thompson L (2004) Mind and heart of the negotiator. Pearson Academic, Upper Saddle River, NJ

Chapter 4
Contract Administration

4.1 Introduction to Contract Administration

The selection of the appropriate contract organization is a central element the owner has to pay attention to during the feasibility phase of a project. A late definition in the contract organization (e.g. when design is underway) may affect the intended results and prevent, for instance, from the usage of integrated-design delivery systems.

Since the contract establishes the rules of the game, the contract itself has to develop according to the project life-cycle: the contracting mechanism has to be chosen during the feasibility stage, the contract documents prepared and finalized during or after design, depending on the delivery system and award method used.

Then the contract has to be managed all throughout the construction phase, and eventually closed-out.

4.2 The Bid and Proposal Management Processes

As discussed in Chap. 3, to award a contract the owner has usually to carry out a bidding process, either as a competition or a negotiation. Differently from private sectors where there are no binding procedures, public award procedures typically have strict law-compliant rules and timing.

With variations depending on local regulations and the nature of the adopted delivery system, typical tasks are as follows.

- Prequalification of bidders, usually based on moral requirements, such as integrity and law compliance, as well as operating capacity and financial reliability. Requirements may be recorded by independent certification authorities.
- Advertising through newspapers and web sites, such as the Dodge Report in the U.S. or the Official European Gazette.
- Mailing or web-based distribution of the bid set of documents usually composed of: 1. Request for Proposal (RFP), 2. Notice Inviting Bids and Instructions to Bidders, which briefly describe the main elements of the scope of work and provides a bid schedule, as well as formal instructions for preparing and submitting proposals; 3. Technical Specifications, Drawings and Plans; 4. Engineer's

A. De Marco, *Project Management for Facility Constructions*,
DOI 10.1007/978-3-642-17092-8_4, © Springer-Verlag Berlin Heidelberg 2011

Detailed Estimates, applicable for DBB contracts; 5. Contract Pro-Forma; and any other forms or templates required (Fisk 2003).

- Opening, evaluation and ranking of all bids that were timely submitted and included complete set of required documentation.
- Public emanation of the Notice of Award.
- Issuance of the Notice to Proceed. In DBB processes, this is done after construction permits and authorizations have been secured.

After an RFP is issued, the contractor prepares the proposal documents looking for the most effective tradeoff between the time to spend for the task and its level of quality, because of the reduced likelihood of getting the contract. Moreover, for DB, turnkey and BOT the proposal process is an expensive and time-consuming activity aimed at producing design, financial and contractual documents that requires multidisciplinary competencies (legal, technical, finance, etc.) and that can be compared to a project by itself. In those cases a proposal management office is needed to direct all proposal projects.

As discussed in Chaps. 2 and 3, the proposal management office (or sales department) may work on a potential project far ahead the RFP, with closed collaboration with the owner for the definition of needs and feasibility studies, especially in the case of CM, DB, turnkey and BOT projects, and make any preliminary decision whether to bid or redirect on better investment opportunities.

By the time an RFP arrives or is advertised, the firm often has appointed a proposal manager, who will prepare a budget and schedule for the proposal project and who is charged with involving and managing the proposal personnel and process. Before the preparation of the documents, the proposal staff may undergo a variety of tasks depending on the complexity of the project and the scope of contract, such as site investigation, evaluation of technological choices, basic design (if applicable), constructability reviews, cost and time estimations, etc. All major decisions affecting the project outcomes are made in the proposal stage.

The results of the proposal management effort are the tender documents, which often specify separate management, technical, and cost-time contents, as well as formal documentation. In the case of a pure competitive bidding based on lump-sum price only, a cost report plus requirement documents are sufficient.

The management contents typically discuss project organization, management methods such as quality and procurement plan, control systems, information tools, etc.

The technical part includes all plans and specs prepared by the contractor, equipment description, as well as options provided for selection by the owner.

The cost proposal contains a detailed price breakdown according to RFP instructions. The breakdown may be either a unit-price quotation of the scope of work or a list of subcontract's prices or both. The cost proposal is integrated with a time schedule, which determines the total duration of the project, and, as a result, the expected cash flow. In some private bidding processes, contractors seek a "present value" of the project return and compute the contract price by discounting the project cash flow (Chap. 7). The bidder's cash flow will allow the owner to estimate the time when payments will occur and, consequently, to determine his own related cash flow.

4.3 The Contract Documents

A contract is typically composed of several documents, as follows.

- The *Agreement*, is a few page signed document that summarizes the main elements of the contract, namely scope, price, baseline schedule.
- The *General Conditions* are usually defined as a standard document that regulates all administrative procedures such as change orders, disputes, etc.
- The *Special Conditions* or Special Provisions add specific project-related issues to the general Conditions.
- *Specifications and Drawings* together form the design documents, either basic or detailed depending on the delivery system. In the first case, the specs describe the functional requirements and expected performance of the project outcomes. Detailed specifications precisely describe all items to be performed with regard to construction techniques to be applied, materials to be used, operating procedures, and other subtleties.
- The *Proposal* or Tender presented by the awarded bidder is usually attached as a formal component of the contract.

The Contract in Practice: Turnkey Contract of a Power Station

Following is an outline sample for a Power Station Contract, where general conditions are edited according to FIDIC guidelines. Typically, the contract documents are divided into three main sets: the contract conditions, the technical specifications, and the turn-key contractor's tender documents that are made part of the agreement.

Section 1 – Conditions of Contract

General Conditions
Definitions and Interpretations
Engineer and Engineer's Representative
Precedence of Documents
Basis of Tender and Contract Price
Changes in Costs
Agreement
Performance Bond of Guarantee
Details Confidential
Notices
Purchaser's General Obligations
Contractor's Obligations

Inspection and Testing of Plant before Delivery
Suspension of Work, Delivery and Erection
Defect before Taking-Over
Variations
Tests on Completion
Taking-Over
Time for Completion and Contract Baseline Schedule
Delay
Performance Test
Defects Liability
Vesting of Plant, and Contractor's Equipment
Schedule of Values, Certificates and Payment
Claims
Patent Rights
Accidents and Damage
Limitations of Liability
Purchaser's Risks
Force Majeure
Insurance
Contractor's Default
Purchaser's Default
Disputes and Arbitration
Sub-Contractors
Applicable Law
Publicity
Special Conditions

Section 2 – Specifications

The Specification
Functional Requirements
Plant Functional Requirements
Site Data
Environmental Constraints
Tests on Completion
Scope of Work
Scope of Supply
Scope of Services
Terminal Points and Interfaces
Contract and Project Management
Documentation Requirements
Training of Purchaser's Staff

> ## Section 3 – The Tender
>
> Scope, Limits and Exclusions from the Supply
> Design Criteria and Reference Conditions
> Plant Layout
> Main Machinery and Systems Functional Descriptions
> Services and Documentation
> Performance Guarantee
> Construction and Commissioning Schedules
> Training

4.4 Contract Bonds

Bonding requirements are usually specified into the contract's general conditions: the contractor has to pay a risk premium to a security company that will reimburse the owner the specific damage if the contractor fails in fulfilling the contract obligations.

Following is a list of the most used kinds of bonds in a construction contract (Fisk 2003). Figure 4.1 shows the different security values with periods of effectiveness.

- Bid Bond: owners request to submit a bid bond as part of the bid documents to protect against the risk the awarder will not get into agreement. In such circumstances, a bid bond typically reimburses the owner with 10% of the bid price estimates. Bid bonds are not expensive: usually insurance companies charge less than 1% of the bid value. The bid bond expires at the effective date of the contract.

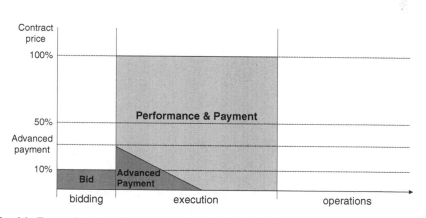

Fig. 4.1 Types of bonds with corresponding stipulated reimbursements to the owner in case of contractor's default

- Performance Bond: the owner requires that the contractor presents a performance bond at the contract signature point in time to shield from incomplete work and unsatisfactory quality, such as might occur in case of insolvency or breach of contract. The bond usually covers the total value of the project to completion with an annual cost between 1 and 5% of the remaining work to do depending on reliability and capacity of the contractor.
- Payment Bond: at signing of agreement, the contractor might also be asked to submit a payment bond. Under this bond, the surety company has an obligation to the owner for the payment of all laborers and suppliers for their legitimate work performed for which the prime contractor failed to pay due to insolvency or other reasons. Payment bonds are usually in an amount of 50–100% of the contract value. A payment bond is a protection to the owner against "Mechanics Liens" that can be claimed by subcontractors and suppliers of building materials who contributed to improvements built on the owner's real estate without compensation. A mechanics lien, if enforced, permits the subcontractor who filed the claim to force a sale of the owner's real estate to pay the claim. From the contractor's point of view, the mechanics lien is the corresponding legal protection from the failure of the owner to pay the contract: because property with a lien on it cannot be easily sold until the lien is paid off, owners have a great incentive to pay their bills (O'Leary and Acret 2001, p. 95).
- Advanced Payment Bond: in case of an anticipated cash the owner requests an advanced payment bond to be reimbursed if the contractor will not perform the corresponding work value. This kind of bond can be issued with a nominal charge. It is largely used in turnkey contracts for the oil, power and plant construction, while very little usage is admitted in public works projects.
- Retained Percentage Bond: owners usually postpone up to 10% of all monthly progress payments 6 or more months after the project is finished to protect against post-construction problems. A *Retained Percentage Bond* may replace the cash retained so as to have the contractor's cash flow improved by total progress cash payments. Under such a bond, surety companies for a nominal charge will grant the owner the reimbursement of 10% of contractor payments if the contractor's work gives rise to defects and liabilities.
- Maintenance (or Mechanical) Bond: this bond may be included in the contract if the contractor is requested to provide a period of maintenance after completion but fails to provide that service.

4.5 Changes and Extra Work

The owner typically can make scope and quality changes throughout the project development by activating a "change order" process to end up with an agreement that reduces, adds to or modifies the work specified out in the contract documents (Fisk 2003, p. 501).

It often happens that also the contractor looks for changes to recover from poor performance; yet, the contract does not always allow for this. Depending on the contract terms, a change might involve additions to or deletion from the scope of work, which in turn affects time and price, alteration of construction methods, materials, and schedule.

A contract change order may be caused by the owner and her A/E, or by the contractor, or externally caused.

The most common sources of owner-generated changes are:

- scope and design changes, sometimes reflecting a desire for the client to redefine the project in light of changing market conditions and needs;
- defects, errors, omissions or ambiguities in the contract original plans and specs;
- unrealistic original quantities, budget or time estimates;
- delayed access to site;
- slow shop drawing submittal approval;
- requests for accelerating the job that require to pay for additional resources and more expensive technologies (see Chap. 8).

The contractor may cause changes because of factors affecting time completion and/or quality of execution, such as a late start in construction operations, use of inadequate resources or poor workmanship, subcontractor/supplier failures, etc.

In some cases, unforeseen external factors may trigger the need for change orders. Unexpected changes to the market and unforeseen site conditions may make original cost or time deadlines impossible to meet. Unforeseen delays may have arisen from changed regulatory issues (zoning, local construction code, environmental constraints, etc.), labor disputes, third party interference, making schedule adjustments unavoidable.

Depending on the type of delivery system and the contract provisions, change orders normally cover direct costs, while the cost of schedule impact must be proven. A contract clause has to specify how the contractor will be compensated for cost and time; it often fails to consider the amplification effects on costs from changing quality, performance, schedule and other factors.

The formal owner-initiated process of change ends up either with a bilateral agreement, referred to as "Change Order", or a unilateral imposition, called "Change Directive".

The Change Order is a formal request by the owner, A/E or CM to add, modify or delete portions of the original scope of work. Here, there is no question that a change occurred, but disagreement may center on the financial compensation for the contract alteration.

If the contractor does not agree to the change request, the owner issues a Change Directive to force the contractor execute the unilateral contract modifications. Usually the contractor works under protest while waiting for a negotiated compensation of work change. Since this approach may be very expensive for the owner or culminate in litigation, an alternate method is to have on-call contractors perform change works under the direct command of the owner.

When a change proposal is claimed by the contractor, after evaluation of A/E or CM, a Constructive Change is agreed between the parties. A constructive change order is usually a major source of dispute because the disagreement centers around the interpretations of contract requirements, plans and technical specifications (Fisk 2003, p. 512). If the approval is unresolved, the change order proposal may escalate to a formal dispute.

4.6 Project Delays

Often, change orders may also be necessary to adjust the original contract schedule because of delays occurred in the project due to unanticipated circumstances. The owner, the contractor or a third party may be responsible for delay.

Under an excusable delay the party is justified from meeting a contract intermediate or final deadline due to external factors that are out of the party's direct control, such as in case of unanticipated weather, labor disputes, or acts of god (force majeure).

If the delay is non-excusable, it is likely that the contractor is asked to shoulder both own and owner's economic consequences. Often non-excusable delays include unavailability of personnel, subcontractor failures, improperly installed work, equipment problems, etc. Non-excusable delays may lead to recession of contract or, more typically, to monetary reimbursement of liquidated damages (Fisk 2003).

The original contract should outline what types of delays are excusable or not, what type are compensable or not, and the impact they may have on the project. Normally, extensions to the contract deadline are not granted and the delay is expected to be absorbed into the schedule. In some situations, disputes between whether a delay was excusable or not, which party should be required to pay for the consequences of the delay, and around the amount of liquidated damages, require formal methods of dispute resolution (Trauner 1993).

Methods such as mediation, arbitration and litigation are available to solve these problems, but the fact remains that formal dispute resolution can only further delay a project. A contractor may choose to continue work under protest, expecting an agreement to be reached at a later point in time.

4.7 Claims and Disputes

Claims and disputes can have major impacts on all aspect of the project, with pervasive influence on the relationships and trust between participants, on progress performance, quality, further delays, morale, and labor atmosphere.

Dispute is a growing problem in the construction industry in most regions of the world, to the point that rarely a project ends with no dispute over differences between the parties. Thus, there is a high need to focus on claim prevention, project management for work to continue under protest, and dispute resolution.

Claims can arise either from the owner or from the contractor under the terms of the construction contract. Claims begin as disagreements between owner and contractor, and often cascade in a disagreement between contractor and its subcontractors and suppliers.

As described, common claim issues include:

- owner-caused delays (e.g. slow review of submittals);
- owner-ordered scheduling changes;
- failure to agree on change order pricing.
- constructive changes;
- nature of differing site conditions;
- bad weather;
- orders to accelerate work;
- loss of productivity;
- suspension of work.

Normally, the contractor must notify the owner of the disagreement. Often this is done through a formal letter of "protest" submitted according to the contract conditions, and to which the owner or the project representative must formally respond. The protest may also be the result of a change order, a change directive from the owner, or an unauthorized change order proposal. If a mutually agreeable course of action cannot be worked out, a formal claim proceeds while work is continued "under protest".

If the parties cannot find a mutually agreeable course of action, sooner or later the conflict has to be solved: during construction or eventually after the project is completed. A claim dispute has usually an escalation: the parties first seek a negotiation; if it is impossible to negotiate, an agreement is attempted by mediation, or, eventually, arbitration or litigation is compulsory.

A systemic contribution to avoid claims is possible (Peña-Mora et al. 2003). First, a responsible owner is required to have up-front clarity of conception, to be a single point of responsibility, to keep an eye on construction (such as in reviews of submittals), and to have realistic buffers in price, schedule, and quality.

A competent and fair project manager or construction manager may help in the task of treating contractors, managing communication among the parties, rapid processing of paperwork, good supervising (via superintendents), careful recording, and proactive detecting and resolving of incipient or realized disputes with minimization of adversarial inclinations.

In-turn, this involves the need for a matching quality design, where plans and specifications are complete, unambiguous, and consistent, where there is coordination of owner & CM & A/E responsibilities, responsiveness of A/E to submittals, monitoring site procedures, and quality inspections of shop drawings.

But overall, this involves a proper contract strategy and design: it is opportune to repeat that the choice of an appropriate delivery system, payment scheme and selection mechanism has key role in preventing a litigious project. Also, the general conditions must contain all risk sharing clause with regard to subsurface conditions, damages due to delays, quantities, change processes, hazardous materials, etc. A contract has to properly address all procedural issues, such as for changes and claims, dispute resolution, payments, etc.

Finally, the following suggestions may help avoid or, at least, reduce claims:

- select a good contractor and have the contractor work with reliable subcontractors;
- develop internal mechanisms to minimize the risk of disputes, namely figure out contingencies that may occur;
- develop mechanisms to allow construction to continue while disputes are being resolved such as using constructive change directives;
- avoid delays in communication;
- confirm all oral agreements/changes in writing and maintain daily records of the project.

4.8 Project Close-Out

The close-out is a major activity of project life-cycle administration, but most often it is miscalculated because managerial attention has been diverted to other projects. Indeed, it is at this point in time that the largest set of contract differences occur, and claims and disputes arise.

Project close-out is not simply a matter of completing the work, having it accepted and finally being paid by the owner. It includes a variety of activities requiring careful attention from both operating and contract perspectives.

The close-out includes all tasks required before the project can be accepted by the owner and the final payment is received by the contractor. This is essentially a small project in itself, sometimes requiring a "close out manager", careful planning and logistics management, but also special attention to emotional issues of the personnel in the project. It presents learning opportunities for the organization through the project final review and report (Meredith and Mantel 2006).

The main tasks in closing out the project are commissioning, termination and feedback learning.

On the one hand, commissioning is about testing and final completion of all work, item and equipment to get to a Certificate of Final Completion. On the other, rigorous contract completion requires all claims are solved, as-built drawings, certificates and paper work are prepared, certificate of final payment is issued, bonds and insurances are ineffective.

The project termination involves operational activities such as demobilization of the construction site, dismantling of temporary facilities and allocation of resources to new projects. Also, termination may include the definition of post-completion guarantees and, especially for plant facilities, of a maintenance obligation to the contractor.

Last but not least, project feed-back includes database updates to benefit future similar projects: standard WBS, list of work packages, unit cost estimates, duration of activities, productivity rates. Also, a company shared discussion of lessons learned from the project allows for method and process improvement.

References and Additional Resources About Contract Administration

Fisk ER (2003) Construction project administration, 7th edn. Pearson Education, Upper Saddle River, NJ

Hendrickson C (2008) Project management for construction, 2nd edn. Carnegie Mellon University, Pittsburgh, PA. http://www.ce.cmu.edu/pmbook/

Meredith JR, Mantel SJ Jr (2006) Project management a managerial approach, 6th edn. Wiley, Hoboken, NJ

O'Leary AF, Acret J (2001) Construction nightmares: jobs from hell and how to avoid them, 2nd edn. BNi Building News

Peña-Mora F, Sosa CE, McCone DS (2003) Introduction to construction dispute resolution. Prentice Hall, Upper Saddle River, NJ

Trauner JT Jr (1993) Managing the construction project: a practical guide for the project manager. Wiley, New York, NY

Part II
Human Resources

A project is basically the implementation of a contract between an owner and a contractor. In turn, a contract is managed by a group of people working for the owner's organization, the contractor's organization, or acting as consultants to one of the parties. To handle and manage the project, those people have to work effectively. This can be achieved by a project-oriented organization, in which responsibilities are defined within project management teams (Chap. 5). Then, to enable project teams work in practice, there is the need for systems and technologies to provide the information infrastructure as well as communications planning and management (Chap. 6).

Chapter 5
Project Management Organization

5.1 The Organizational Challenge

The contract is the core of a project because it defines roles and responsibilities of the parties, financial relationships and rules for project execution.

But, overall, human resources are key factors to implement and manage the contract: operational personnel have to execute the project while the management staff has to make decisions, plan and schedule, monitor performance, communicate and circulate reports, direct, and control the project.

Even given the same contract, the cohesiveness of the people on a team and the quality and effectiveness of their interactions can make the difference between a project that is a great success and one that develops into a disaster.

Persons involved in the project effort act differently according to the role assigned by both the contract mechanism and the project organization. In small organizations it may happen that a single person wears different hats, such as in one-man organizations.

More often, as firms grow, adding projects and human resources, the organizations tend to be more and more sophisticated and complex, developing an organizational structure (Meredith and Mantel 2006). Moreover, organizations have to conform to diverse types of projects, various contract agreements and different management staff and practices.

There are several types of project-oriented organizational structures. Many of the differences reflect the function the firm has in the construction industry and the orientation to project management. For example, as shown in Fig. 5.1, the same project may have three different project management organizations working together (here, the owner project management team, the contractor project team, and the construction manager's staff), which are in turn comprised into the organizational structures of the parent firms.

However, because organizational principles exhibit minor variations across many companies, commonly-used organizational structures are rather few. A key challenge is to develop and maintain the most effective organizational structure for supporting projects teams, maximizing coordination and sustaining the work environment.

A. De Marco, *Project Management for Facility Constructions*,
DOI 10.1007/978-3-642-17092-8_5, © Springer-Verlag Berlin Heidelberg 2011

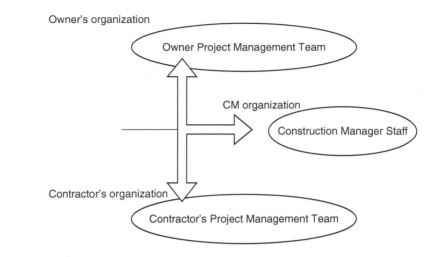

Fig. 5.1 Project teams and parent organizations

5.2 Organizing the Firm for Project Management

Figure 5.2 presents the main organizational structures according to the level of interface between the project and its parent organization.

Traditionally, in functional organizations there are no project managers, while a task-force can be considered as a temporary child firm dedicated to carrying out a specific project assigned to a responsible Project Manager acting as its C.E.O.

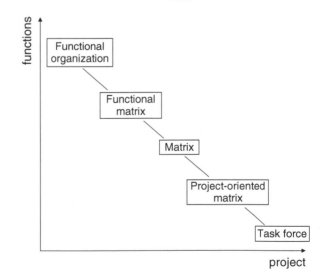

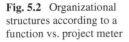

Fig. 5.2 Organizational
structures according to a
function vs. project meter

A balanced point between functional and task-force models is represented by the matrix organization where functional managers and project managers coexist, with differences in function or project orientation.

A functionally-organized construction firm is divided into functions covering every domain of the firm. Each function has its own functional or line manager. All projects are decomposed into major work packages, one for each interested functional discipline.

The line managers together with the executive office are responsible for project management. Typically, a functional contracting firm has an organizational chart similar to the one illustrated in Fig. 5.3.

Accordingly, the engineering division is likely to have a vice-president controlling several discipline managers, such as the architectural design manager, the manager for civil engineering, the one for mechanical engineering, electrical, and so on.

The advantages of a traditional functional structure are concentrated in three main areas. First, technical know-how is maximized within the line-specific division where knowledge and experience relevant to that discipline are shared among specialists, and where technological, procedural and administrative continuity is assured.

Second, personnel have substantial flexibility to work on different projects at once. Finally, budget responsibilities are clearly partitioned within each division and the top management is guaranteed a centralized control.

Yet, the functional structure lacks cross-functional horizontal coordination and integration of all components and interconnected competences required to execute a project. This results in poor client orientation, hierarchical communication and fragmentation of responsibilities across the firm. Also, human resources tend to specialize in their specific technical expertise far more than develop managerial skills to face risk and uncertainty.

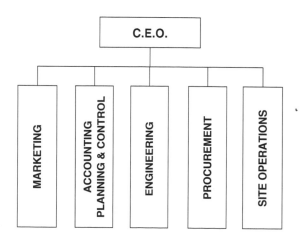

Fig. 5.3 Example of a functional organization chart

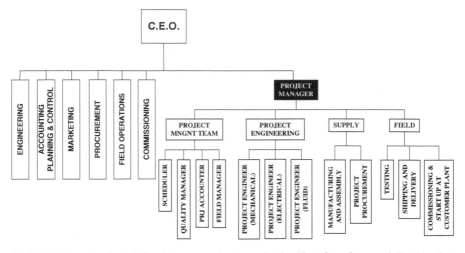

Fig. 5.4 Example of a task-force or pure-project organization chart for a firm specialized in plant engineering

Opposite to functional organization, a pure project task-force is used to execute a unique and important project, as a child separate company (Fig. 5.4). A project manager is appointed to be fully responsible for the project development with the assistance of a full-time project team.

Pure project organizations are commonly used by general contractors to staff construction sites far-off the headquarters.

A task-force has a central advantage: authority and direction for a project are centralized, and the structure facilitates direct communication and systems integration. The organizational structure is flat and informal, enabling project commitment to all participants, a friendly work environment and tight relationships between all the parties involved. Conversely, the most important disadvantages are mostly associated with the cost of overstaffing (mostly reflecting the fact that a task-force requires full-time personnel dedicated to the project), logistics and duplication of services still existing in the parent company.

To overcome the limitations of both models, the matrix organization attempts to combine the functional structure with the task-force one.

Within a matrix organization, project teams are not separated from the parent organization and are instead managed by both the project manager and the line managers with regard to different tasks. The project manager is responsible for project planning and scheduling, while line managers organize know-how and make personnel available to the project.

Budgeting is a shared activity involving both scope planning and resource allocation. Briefly, cost and time are handled by the project manager, while functional managers cope with people and quality.

Human resources work in the functional division and are "lent" to the project manager for the required amount of working hours, as shown in Fig. 5.5. Specialists

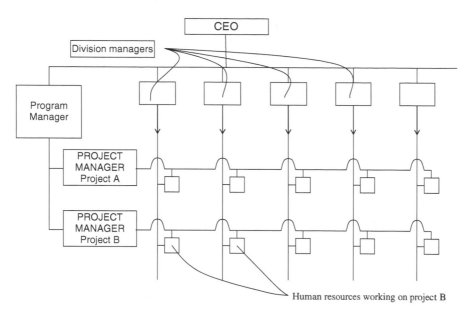

Fig. 5.5 Example of a matrix organization chart

have the necessary knowhow and tools available in their functions, and share their working time between several projects.

As a combination of task-force and functional organization, the matrix can assume different forms in the organizational spectrum (Meredith and Mantel 2006); these primarily reflect differences in the project manager's responsibilities and authority.

In a "functional matrix" the line managers are responsible for carrying out the project with the help of a Project Coordinator enabling cross-functional project communication and integration. The project coordinator is in charge of coordinating the scope of work performed by the different divisions.

In a "balanced matrix" the project manager and the division managers have balanced authority on the project budget, with the project manager acting more as a planner and controller, while line managers manage the technical aspects and the people's effort.

In a "project-oriented matrix", the functional managers only provide know-how and tools, while project managers cope with everything else.

Matrixes are typical in firms that sell project management services, such as construction management firms and design-builders, where system integration, fast-track and construction speed require effective coordination. Unfortunately, the matrix organization often results in a persistent competition between division and project managers, as well as in disputes over responsibility for missing goals.

It is obvious that a matrix organization cannot be installed or simply implemented in a company traditionally working with a functional organization or by

pure projects. Mixing the two approaches requires a strong management commitment to face change resistance and takes a long time because processes have to be
reengineered and people educated in a project management perspective (Archibald
2003).

The challenge is much easier in the case of partnerships and new consortia created with the special purpose of carrying out one single project: two or more firms
come together to bid for a particular project (a temporary joint-venture) or to secure,
for example, better chances in a new market (a stable consortium).

A key factor here is to provide the project manager with sufficient authority to
make all people from all companies work for reaching a successful completion of
the project and not just for the profitability of a single partnering company. As a
result, a project-oriented matrix is established, with each organization acting as a
functional division providing services to the joint project company.

5.3 Organizing the Project Team

The number of people to be hired in a Project Management Team usually increases
with project size and complexity.

The Project Manager (PM) with an assistant, both working on two or more
projects at once, would probably be sufficient to carry out small projects.

In large projects, different professionals are needed to support the PM in performing various tasks. Key activities of the PM are, first, to design the project team;
second, to get the right persons for the various jobs, and, finally, to build up and
motivate the team to enhance project performance (Barrie et al. 1992).

Typically, the roles that join a large project team as follows.

- The Project Engineer is responsible for design; she is in charge of coordinating
  and integrating design contributions from specialty designers (such as architects,
  mechanical, electrical engineers, etc.), for constructability and design changes.
- The Project Planner supports the PM in planning and scheduling the scope
  of work.
- The Accounting Manager helps define the budget, monitor the project performance, keep track of cost consumption during the project execution, and prepare
  financial reports.
- The Procurement Manager is in charge of all purchases of construction materials
  and services.
- The Quality Manager has to establish a correct quality environment and control
  quality assurance procedures to make sure the constructed facility will achieve
  the specified level of quality.
- The Risk Manager contributes in planning, monitoring and controlling risk and
  uncertainty of the project.
- The Site or Field Engineer is responsible for construction, installation and testing
  of the facility. She is often helped by one or more Construction Superintendents.

Where agreed upon in the contract (such as in most DBB projects), the owner may appoint a Resident Project Representative or Resident Engineer to keep an eye on the construction site and to work closely with the contractor's project management team.

The design of a Project Organization Chart (POC) helps the project participants understand their roles, as well as the hierarchic relationships between them, as in the example shown in Fig. 5.6.

Usually the project team is part of the overall project Organizational Breakdown Structure (OBS), which also includes all the other participants such as owner, contractors, external designers and suppliers, etc. The aim of the OBS is to list the names of resources required and to assign the right job to the right person, as it will be better discussed in Chap. 11.

As discussed in Chap. 3, a centralized Project Management Office (PMO) is often used to support project teams. The PMO provides services that are common to all projects, such as:

- bidding (in substitution of or together with the sales division);
- resource pool management,
- reporting project progress and evaluating performance,
- defining standards, processes and tools to enhance the project management practice within the enterprise,
- maintaining and updating the information database.

The role of the PMO in a company if often crucial to assure project management organizational maturity and the process of continuously implement and improve PM practices and standards.

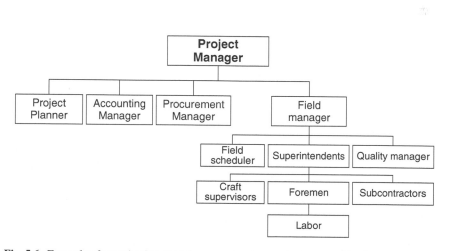

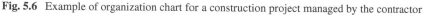

Fig. 5.6 Example of organization chart for a construction project managed by the contractor

PM Standards and Certification Programs

Here is a list of the most spread and globally recognized non-for-profit associations that serve member practitioners and organizations with project management standards, good practices, resources and certification programs.

- The Project Management Institute (PMI®, www.pmi.org) issues a quantity of resources for project management. In particular, A Guide to the Project Management Body of Knowledge (PMBOK® Guide) is considered as a reference global standard, together with the practice standards for PM processes and the Program and Portfolio Management Standards. PMI also offers a certification program at various competence levels for project practitioners of all industry. Moreover, a specific recommendable standard has been developed for the construction and building sectors: Construction Extension to the PMBOK® Guide.
- The International Project Management Association (IPMA®, www.ipma.ch) also offers a four level certification program based on the IPMA Competence Baseline, which sets out the knowledge and experience expected from managers of projects, programs and project portfolios.
- PRINCE2® (PRojects IN Controlled Environments) is a process-based method originating in the UK government, used in both the private and public sector. PRINCE2® is composed of free resources registered under the trademark of the UK Office of Government Commerce.
- The Association for Project Management (APM www.apm.org.uk) develops the APM Body of Knowledge and provides a knowledge based qualification program, in integration with PRINCE2® methodology.

5.4 People and the Project Manager

The role of the PM is to make decisions, plan and control uncertainty to bring the project the most as possible in line with the expected cost, time and quality, or, at least, to find out ways to reduce time delays, cost overruns and quality cutback.

But, overall, the PM's first aim and concern is to direct all people involved in the project challenge. People may include communities of users, external and internal stakeholders, owners, internal project team and labor, as well as external designers, sub/contractors, suppliers and consultants.

To this end, a PM is not only required to plan, schedule and control, but also to have comprehensive experience, knowledge of human behaviors, psychology, leadership attitudes, communication skills, and multidisciplinary knowledge in several areas, such as technology, organization, finance, legal and contracts, administration, negotiation techniques, etc.

In particular, the PM has to exercise authority, decision, and control within the boundaries of the contract arrangements that are set between all the parties involved in the project.

When dealing with external parties and stakeholders – who are often central to the success of a project – the PM would need to take advantage by adding good human communication practices to proper contract administration.

In internal relationships, it is crucial that the PM acts in a double way: on the one hand, she would establish virtual internal contracts that assign precise roles and charge each team member with strict responsibilities; on the other, she would motivate individuals and build up the team.

As a consequence, individuals would be willing to effectively do their assigned tasks within the time allotted, and quickly report unanticipated problems, resulting in a successful whole project.

A good project management practice contributes to establish a collaborative environment between people involved in the project with different roles and responsibilities. The mechanism is presented in Fig. 5.7.

As people are naturally keen to maximize individual utility, also companies look for their own return. Project managers have to take this into account and have each resource, either a person or an entity, work on a project with a specified role and contract responsibility to complete an assigned task within the allotted budget and time, and give this resource enough incentive to get a satisfactory return.

Leveraging individual responsibilities is a good manner to establish cooperation, coordinate and manage communication between entities that are likely to work to meet deadlines and budgets, which in turn affect a sequenced multi-party construction job.

Such an approach seems to be particularly valuable for international projects, where various organizations and staff join from a diverse set of backgrounds and places of origin.

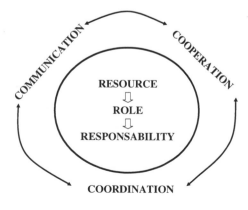

Fig. 5.7 Key concepts to enable project collaboration: "3R" management of personnel can establish a "3C" project environment

In such projects, the PM would focus on intercultural management to reconcile cultural and operational differences of individuals and of the groups they belong to: when defining roles, rules and tasks she would take into account their individual and corporate approaches to, for example, relationship, team work, and communication, trying to maximize everyone's attitudes and values (Trompenaars and Hampden-Turner 1998). Key aims are to motivate the team and to solve any possible conflicts, and this can be done by using effective communication.

References and Additional Resources About Project Management Organization

Archibald RD (2003) Managing high-technology programs and projects, 3rd edn. Wiley, Hoboken, NJ

Barrie DS, Paulson BC, Paulson B (1992) Professional construction management, 3rd edn. McGraw-Hill, New York, NY

Meredith JR, Mantel SJ Jr (2006) Project management a managerial approach, 6th edn. Wiley, Hoboken, NJ

Trompenaars F, Hampden-Turner C (1998) Riding the waves of culture, 2nd edn. McGraw-Hill, New York, NY

Chapter 6
Project Information and Communications Management

6.1 Role of Information and Communications

An organization is based on communications as far as it is the one and only information driver among people involved. A key factor for success in this area is to have the information correctly generated as a document or a message – whether oral or written – and then distributed according to processes and media that are effective to the organization. Accordingly, efficient communication processes are necessary to enhance project performance.

Today, information is increasingly generated in digital form. Data are collected from an electronic filing system and information is produced using a number of software packages for a variety of activities.

Yet, the communication process is often not effective and poor cross-discipline and inter-actor communication is one of the major bottlenecks to project performance improvement (Sun and Aouad 2000).

The reasons may be mainly drawn in two areas: technology and communications management. First, electronic connections between different firms working on the same project, as well as interoperability standards between different software packages are not common or are hard to handle in an integrated framework. In addition, the selection of useful software for information exchange to support project management is not easy, because of its rapid evolution, wide market offering and high costs (Egan 1998).

Second, communication processes may be not correctly designed and may generate time-cost consuming flows. The definition from the planning phase of proper workflows will help in improving efficiency. But more often, project teams do not have real-time access to up-to-date information resulting in faulty managerial actions based on perceived performance rather than on actual one.

6.2 Technologies and Systems for Project Management

Project teams and other project participants have to work with common standards, share documents, gain access to timely information, be notified of events: briefly, effectively communicate. The availability of this information serves

A. De Marco, *Project Management for Facility Constructions*,
DOI 10.1007/978-3-642-17092-8_6, © Springer-Verlag Berlin Heidelberg 2011

two broad purposes, namely: facilitating operations and improving decision making.

The effectiveness of the communication process is obviously the result of the way people interact and work, but it is strongly enhanced by a proper set, and usage, of information systems and technologies, which help create, share and manage knowledge (Anumba et al. 2004).

A set of different hardware and software tools is needed, both for the overall organization and for single projects. Four main components should be available in a construction firm: a filing system, a set of specialized software packages for individual productivity, an scheduling/accounting tool for project planning, and a collaborative workplace.

6.2.1 Filing System

A filing system such a data warehouse or a simple file repository should be accessible via the web to enable staff, suppliers and owners to access information from distributed locations. Entry and access to particular information should be restricted according to the user profile and the specific role in the project workplace.

6.2.2 Individual Productivity Tools

Specialized tools for individual operations help design, engineering, estimating, reporting and decision making activities. Data collection for operations or management is made from databases, data warehouses and filing system.

The production of information is supported by a number of software packages for a variety of activities, such as back-office suites, Computer Aided Design (CAD) and Building Information Modeling (BIM) software tools, engineering and design calculators, decision support systems, tools for editing procurement, safety and quality documentation, etc.

6.2.3 Project Planning Tool

A Project Planner is a software application that allows detailed project planning, scheduling and monitoring processes. It usually contains functionality for optimization of schedules, costs and resource usage. The project planner should be integrated with the corporate bookkeeping and accounting system, which is usually referred to as an Enterprise Resource Planner (ERP). The ERP supports multiple project management and a proper resource allocation process.

> ## Project Planning Software Tools
>
> Here are just some of the number of project planning software tools available on the marketplace. Most PM software manufacturers produce various corporate solutions for both project and portfolio management, and activate both industry-specific and customized solutions.
>
> - Artemis®
> - Microsoft® Project
> - OpenProj
> - Oracle's Primavera® Project Planner
> - Siemens Team Center®

6.2.4 Collaborative Workplace

A collaborative workplace is usually defined as an environment where people can communicate, exchange notes, and manage information processes. It may be defined as a database-backed mechanism that allows real-time notification of changed information to all interested parties, document workflows, as well as tracking of all communication between the participants, thus enabling a collaborative environment. It typically uses a web based access and also may use wireless systems to connect suppliers, owners, contractors in a sole communication standard.

Mainly, the core function of a collaborative workplace is the document management system, which may be enhanced by communication functionalities such as electronic messaging and workflow settings for submission, revision and approval of documents.

The connection between the project planner and the document management tool provides availability of data and documentation that is needed to perform an assigned task, or simply brings automation to all planned communication processes, according to a scheme such as the one in Fig. 6.1.

The more the document management system contains added tools and integrates with the other systems, the greater the monitoring and control capabilities over single projects and the portfolio of projects. Thus, the document management system becomes a supporting control system for progress and performance monitoring, as shown in Fig. 6.2.

6.3 Communications Management

Communications management requires to plan and maintain the process of assuring timely and appropriate distribution and sharing of project information among stakeholders and all actors involved in the project organization. The challenge here is to

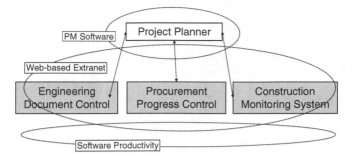

Fig. 6.1 Example of a document management system for engineering, procurement and construction projects (EPC)

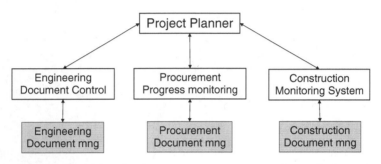

Fig. 6.2 Example of a collaborative document management system for monitoring and control of EPC projects

drive the right piece of information to the right person at the right time, so that the entire project team is given a framework for effective and profitable collaboration.

Information systems help in the task of generating, collecting and storing data, while software applications for communications management enable a timely distribution and dissemination of information. In other words, information technologies are the supporting backbone for a valuable communication management process, which is planned and then developed during the project life-cycle.

A communications plan (Project Management Institute 2008) is prepared to determine:

- the technologies and standards to be used in the project for generating, filing, and distributing documents;
- objectives, priorities and constraints in the record keeping and communication process;
- the kinds of documents, messages, reports and presentations that will be generated and distributed, including templates for all types of report, as well as the information they have to contain;
- the responsibilities of addressees of the workflow as part of a review and approval procedure.

During the design phase, drawing/design reviews, meeting minutes and engineering progress reports are some of the most common communication artifacts that are disseminated among project participants.

On the job site, project plans and schedules, construction progress reports, job conferences, inspection, quality and safety reports, construction diaries, progress reports, requests for information (RFIs), punch-lists, contractor submittals, change orders and directives are used on a daily basis to manage communication between the parties.

From the management point of view, periodical reporting weekly or monthly disseminates information about project status, schedule and cost records, cash flow tracking, performance analysis and forecasting, as presented in the next section.

References and Additional Resources About Information and Communication Management

Anumba C, Egbu C, Carrillo P (2004) Knowledge management in construction. Blackwell Publishing, Oxford, UK

Egan J (1998) Rethinking construction. Department of Environment, Transport and Regions, London, UK

Enshassi A (1996) A monitoring and controlling system in managing infrastructure projects. Build Res Infor 24(3):183–189

Fisk ER (2003) Construction project administration, 7th edn. Pearson Education, Upper Saddle River, NJ

Project Management Institute (2008) A guide to the project management body of knowledge, 4th edn. Project Management Institute, Newtown Square, PA

Sun M, Aouad G (2000) Integration technologies to support organizational changes in the construction industry. ISPE Int Conf Concur Eng 7:596–604

Part III
Money

Money, which is here referred to as cash made available to sustain a capital investment, is a major concern throughout the life-cycle of a construction project. During the feasibility stage (Chap. 7), the decision to proceed is carefully made based on evaluation of project profitability. Also, funding opportunities need to be investigated and appropriate shares of equity and debt funds have to be determined into the capital structure. During the planning phases (Chap. 8), budgets and timeline schedules are prepared using dedicated techniques and tools. The planning activity allows for forecasting cash streams. Then, the development phase requires that the project is periodically monitored and controlled (Chap. 9). This requires that a set of progress measurement activities are established at the project management level to support the process of continuously estimating the actual completion time and final cost and to help making corrective actions to bring the project in line with the initial plans.

Chapter 7
Project Feasibility

7.1 Project Financial Engineering

Undertaking evaluation of a project feasibility essentially means to pursue a project idea by supporting investigations on several components such as land purchase and sale review, constraint survey, cost/revenue models, permit requirements, risk analysis, etc. All these assessments during the feasibility phase help owners and contractors to evaluate the project and to make financing plans as precursors to any decision to proceed with the project execution.

The development of new construction projects often entails high costs in the short term while benefits only accrue over the long term. Since costs occur prior to income, capital funds are required to face life-cycle financing (Hendrickson 2008).

For owners and real estate developers, project financing attempts to bridge the gap between short-range expenses, due to contractors for their construction job, and long-term revenue from operations, and to solve the problem of negative balance of the project cash flow.

Figure 7.1 shows typical cumulative cash flows in the case of a traditional design-bid-build DBB contract. In this case, the owner's investment is reimbursed over a long time frame, while the contractor has to advance a smaller cost for a shorter period of time.

Financing is absolutely crucial to making capital projects possible. The possibility of funding a project is the prerequisite for its initiation. Furthermore, even when financing is secured, it is a major driver towards alternate delivery methods, and the financing mechanism has an impact on risk of construction and often significantly influences who shoulders the risk of the project. It also influences the type of construction undertaken, bid prices offered by contractors, and construction claims.

7.1.1 Owner Financing

With most construction delivery systems, owners are in part or totally responsible for securing capital for the implementation of the project. Methods of financing differentiate depending on the public or private nature of the owner and on the private or public destination of the project.

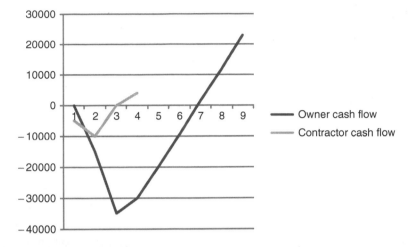

Fig. 7.1 Cumulative cash flows of a traditional DBB contract

To fund public works and specific public facility projects, public owners might make use of tax revenues, bonds, capital grants subsidies, and international subsidized loans.

With a general purpose bond, a government redeems future general taxes to pay back the principal plus additional interest at a future time. Special-purpose bonds are repaid through taxes collected from the people benefiting from the project, or from user fees collected for the project. Examples of this type of bond are those used to fund toll roads. It should be noted that an advantage of public financing is that government bonds are usually tax-exempt. This makes finance much cheaper for public institutions. However, public owners face restrictions such as debt limits, bond caps or lack of support for bonds. Many bonds require the people involved to vote for its approval. If it is not approved, public institutions may lack the necessary funding to go through with their project. This is one of the major motivations for the formation of public/private partnerships (e.g.: Build-Operate-Transfer). In this situation, the private partner shoulders some of the cost.

Public institutions can also request federal, state or city grants to fund their projects. This is often used to fund highway or public transportation projects. In developing countries, funds for infrastructure projects can also be obtained through internationally subsidized loans. The World Bank, Asian Development Bank, European Bank for Reconstruction and Development and other international development institutions are the largest providers of funds for the purpose of development.

Since the sources of public funding are generally derived from the citizens, public projects must demonstrate that they offer important social benefits. This could mean user surplus, regional growth, unemployment relief, poverty alleviation, etc. Since many of these benefits are not easily quantifiable, they are generally omitted from the financial balance sheet for a project. This is a significant reason that, the Minimum Acceptable Rate of Return (MARR) on government investments is generally much lower than that of privately funded projects.

Private financing is secured by equity capital (corporate resources, or multiple-corporate resources in case of joint ventures) and debt capital (in the form of construction loans, long term mortgages, or leasing). Major mechanisms are based on financial leverage: part of the investment is covered by corporate equity (by direct investment of corporate capital and retained earnings, by offering equity shares and stock issuance, e.g. in capital markets) and part of the investment is borrowed from either investors (bonds) or lending institutions (loans and mortgages).

The use of debt capital involves borrowing money from a bank or other financial institution in the form of a loan. In the particular case of a collateral constructed facility, a short-term loan bridging construction and start-up periods is converted into a long-term senior mortgage reimbursed by operations. The construction loan is a high-risk loan and hence has high interest rates. Once a completed facility can be seen, lenders are willing to provide lower interest loans since the risk is lowered and secured by the collateral asset. The lender of the long-term loan may or may not be the same as the lender of the construction loan. As a rule of thumb, such a loan must usually be repaid by two thirds of the total project projected life. Repayment is secured through project revenue or other sources of income.

In most cases, the capital provided by the lending institutions is less than the cost of the entire project. This difference – which helps lower the risk of the project to the lender – constitutes the owner's equity, and must be provided from the owner's own funds. Although there is no explicit cost for the use of such capital, this option will not be attractive to an owner unless the project has a high enough expected rate of return. This is a result of the opportunity cost of capital, the fact that investing this capital in this project prevents its use in other profitable investments (Sullivan et al., 2001).

Due to these considerations, companies most often use a combination of equity and debt to fund their infrastructure projects, within the ceiling loan size determined by the maximum level that can be paid off by the anticipated income of the project: the maximum loan amount is the yearly net income divided by the capitalization rate (Formula 7.1). In this case, the capitalization rate is the weighted average of both lender and borrower's interest rate (the Weighted Average Cost of Capital).

$$loan\,amount = \frac{\text{average yearly net income}}{\text{WACC}} \qquad (7.1)$$

The WACC is calculated summing the weighted averages of a company's cost of equity and cost of debt:

$$WACC = \frac{E}{E+D} \times Ce + \frac{D}{E+D} \times Cd \times (1 - Tc) \qquad (7.2)$$

Where:

- Ce = cost of equity, referred to as the percentage expected rate of return on equity, were it invested in other corporate projects.
- Cd = cost of debt, referred to as the interest rate of the loan
- E = Total equity investment
- D = Debt, as the loan amount
- Tc = corporate tax rate (if interest is tax deductible)

Following is a simple example of how this can be used.

Example – Computing the Loan Amount

Suppose a firm is willing to construct a new factory that will cost $15 million, and is anticipated to return a net income of $1 million per year.

The firm is able to secure a loan at an interest rate of 7.5% and its cost of equity is 10%. Interests are not tax-deductible.

To solve this, we use the formula (7.2), as well as our knowledge that the debt amount D is equal to net income/WACC.

We can now substitute $\frac{income}{WACC}$ for D, and the balance of the cost $\left(\text{project cost} - \frac{income}{WACC}\right)$ for E.

We then use an initial estimate value of WACC (e.g. 10%) on the right side of the equation, and iterate to find a solution.

We find WACC = 7.9%. Therefore, the firm will take out a loan of $12.7 million and provide the remaining $2.3 million using its equity.

Of course, the bank may not want to lend all the $12.7 million to the firm and would seek a transaction that represents a likely valuable investment for both parties. In general, when faced with the opportunity to lend capital to a firm, a lending institution will look at the character of the firm, the firm's ability to repay, and the existence of collateral assets. The character of the firm is its attitude towards repayment. This is usually historical data, and tells whether and how quickly the firm usually pays back its debt. The firm's ability to repay can be analyzed using various documents such as financial statements from the owner, land titles, proof of zoning, retained earnings accounts reconciliation, design documents, cost estimates for design, market research for estimating demand, detail pro-forma and other financial documents. Often, lending contracts will also include collateral agreements, where a firm's assets can be taken in case the project cannot repay its debt.

As an alternative source of funding, owners may lease a constructed facility from a developer or enter into a BOT kind of contract where funding is engineered using a "Project Financing" perspective.

7.1.2 Project Financing

Project financing involves the creation of a legally independent Special Purpose Entity (SPE) or Vehicle (SPV) company for the purpose of investing in a single purpose industrial asset or constructed facility and of segregating the associated cash flow and risks from the shareholders. In fact, the SPE helps to shield the investing companies from the project financial risks through nonrecourse debt. With nonrecourse financing lenders have no recourse for repayment of their loans against the shareholders, but only against the SPV segregated cash flows and assets (Finnerty 2007).

These SPEs are usually owned by several companies in a joint-venture, although a single company often owns a majority of the SPE.

Major examples of this type of financing are the Eurostar tunnel under the English Channel, EuroDisney, or the Bangkok elevated road highways.

In this type of financing, the ratio of debt to equity used is dependent on the project's ability to reimburse its loan every year. The project capacity to reimburse its debt is measured by the Debt Service Coverage Ratio (DSCR):

$$D.S.C.R = \frac{\text{Annual Project Cashflow}}{\text{Annual Repayment Amont} + \text{Interest}} \qquad (7.3)$$

As a rule of thumb, the minimum pre-tax DSCR for any given year should not be below 1.2, while the average should be 1.5. Minimum post-tax DSCR should not be below 1, with an average of 1.2. Yet, the indicated DSCR values may vary depending on the risk profile of the project and the shareholding companies.

If on a given year, post-tax DSCR is below 1, then the project cannot meet its debt servicing obligations for that year, and arrangements must be made to avoid that. Therefore the amount of debt is determined so as to keep the DSCR within healthy limits. The remaining funds arc obtained from the equity of the parent companies. The portion of equity that provides such a healthy balance is usually about 15–30% of the total financing needs.

7.1.3 Contractor Financing

From the point of view of the contractor or subcontractor, the cash-out is typically a S-shaped curve line (so-called "S-curve", see Chap. 8) as a result of cumulative labor and material cost payments required to perform the project scope of work, while cash-in is related to the terms of payment specified in the contract.

There are many different terms of payment, the most common being a monthly reimbursement of work completed ("Schedule of Values"). The expenses incurred by the contractor form a relatively continuous curve. The contractor receives payments from the owner either on a time basis or milestone basis, both represented by step functions. These payments lag behind the expenses, which often puts pressure on the contractor's cash flow.

Furthermore, as a risk-reducing mechanism and to provide incentives to the contractor for timely completion, the owner usually retains some payment (usually around 10%), with the retained amount only flowing to the contractor at the completion of the project.

Figure 7.2 shows the typical cash flow of a contractor resulting from monthly progress reimbursements. The negative balance is secured by a bank overdraft which includes interest charges.

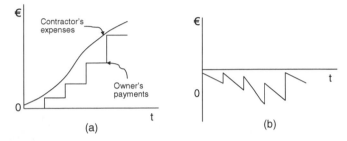

Fig. 7.2 (**a**) Contractor's expense curve and owner's payments. (**b**) Contractors' overdraft (Hendrickson 2008)

Due to payment delays and retentions, contractors have to arrange for bank account overdrafts as a cash anticipation of future contracted incomes. In some projects, a three-way payment agreement is made between a contractor, an owner and a bank. In this situation, the contractor submits a monthly progress report to the owner, who then submits a "draw request" to the bank (Hendrickson 2008).

In any case, a framework of payment must be agreed in the contract. The structure of payment is usually proposed by the owner. In a responsible project, the owner should check progress and cost reports submitted by the contractor. For further details see Chaps. 8 and 9.

7.2 Financial Evaluation of Projects

As part of the Program Management process, Chap. 1 discussed about several qualitative techniques that can be used to evaluate and rank different investment opportunities, such as SWOT analysis, as well as some semi-quantitative models based on comparative benefit techniques and weighted multiple criteria.

Qualitative methods are rarely sufficient to evaluate a given investment or project. Adjective methods scoring weighted categories are often useful for choosing between similar projects or proposals, but fall far short in financial evaluation. Therefore, quantitative models are often employed to better understand the value of investment opportunities.

7.2.1 Net Present Value

Quantitative evaluation of projects is based on the "time value of money" concept.

If one assumes that money today can be invested in a reliable source, such as a bank or a government security, then that money will produce some future gain with interest.

Thus, by investing some amount of money today, a larger future amount can always be produced; money in the present has an equal worth to a larger future value of money.

Example – Future Value

Suppose that a reliable lending institution offers an annual interest rate of 5%. Investing $100,000 today would result in $100,000 * 1.05 = $105,000 1 year from now. At this interest rate, the 1-year future value of $100,000 today is therefore $105,000.

More generally, for a given annual interest rate, r, the future value, FV, of V dollars today at a time t years into the future is simply given by:

$$FV(t) = V * (1 + r)^t \qquad (7.4)$$

Similarly, future money is worth a lesser "present value" of money today: the far a given sum of money is into the future, the smaller its present value today. Rearranging (7.4) produces a simple formula for calculating the present value at time 0, PV(0), of a cash flow of V dollars occurring at time t in the future:

$$PV(0) = V/(1 + r)^t \qquad (7.5)$$

The term $\frac{1}{(1+r)^t}$ that converts a future value into a present value is referred to as the *discount factor*.

The present value can be thought of as "equivalent" to the future value in the sense that we could convert one into the other without any cost.

- To convert a present value (in hand at the present) into its equivalent future value at time t, one merely has to invest the present value into the reliable investment vehicle, and withdraw the money – with accrued interest – at time t. The size of the resulting withdrawal (principal plus interest) will be exactly equal to the anticipated future value.
- To convert an anticipated future value to be received at time t into its present value equivalent in hand at present, one can simply borrow the present value money from the reliable source. At time t, the amount of money owed the reliable source is exactly the future value at time t; this balance can be completely paid off by the value received at time t.

Assuming that investors are rational decision makers, it can be implied that an investment will not be undertaken if its expected return is less than the one offered by the reliable source. If more than one reliable investment opportunity exists, an investor would be indifferent between future amounts of money, or cash flows, producing the same present value. If the present values of two or more investments are the same, the investments are of equal financial worth today.

Example – Comparing Present Values

Assume that a same initial sum of money might produce two different investment opportunities: investment A has a cash flow of $100,000 1 year from now while investment B has a cash flow of $105,000 2 years from now.

If a discount rate $r = 5\%$ annually is assumed, even though the nominal cash flow of B is $5,000 greater than that of A, both investments have the same present value and are therefore of equal worth today.

$$PV_A = \$100,000/(1 + 0.05)^1 = \$95,238$$
$$PV_B = \$105,000/(1 + 0.05)^2 = \$95,238$$

Had the discount rate been different than 5%, the present values of investments A and B would not have been of equal worth: a lower rate would have resulted in B being a better investment than A ($PV_A < PV_B$), while a higher discount rate would have suggested that A is a more valuable investment than B.

Projects are investment opportunities involving a stream of positive cash inflows and negative outflows over the project duration. The net present value (NPV) of the investment results from discounting the cash flow. Revenues or expenditures that occur today are included in the NPV summation at their nominal values (7.5).

$$PV = V/(1 + r)^0 = V \tag{7.5}$$

Choosing the discount rate becomes important in the accurate calculation of NPVs, as it must capture the risk of an investment and the opportunity cost of not investing in some other opportunity. A more detailed discussion regarding the choice of an appropriate discount rate is included in a later section. For the time being, it is acceptable to simply assume the discount rate is equal to the interest rate available from a reliable source. With this assumption, the NPV of a project becomes the value of the project's cash stream (in present value terms) beyond what could be gained from investing in the reliable source.

The NPV of investing in the reliable source itself would therefore be zero. Projects with an NPV greater than zero would be more valuable than investing in the reliable source, and projects with a negative NPV would be less valuable than investing in the reliable source.

Thus, NPV is an effective method for deciding whether or not to undertake a given investment opportunity.

Example – Net Present Value

Consider a reliable source with a 10% annual interest rate. Suppose $100,000 is invested in this reliable source today, future interest earnings of $10,000 per year (received at the end of each year) are reinvested at the same rate for the next 3 years, and the original $100,000 investment is returned at the end of year 3.

The NPV of this investment is simply:

$$NPV = -100,000 + 10,000/(1+0.1)^1 + 10,000/(1+0.1)^2 + 110,000/(1+0.1)^3$$
$$= 0$$

The opportunity cost of not investing in the reliable source is illustrated by the following example.

Assume the reliable source with 10% annual interest rate still exists. However, instead of investing $100,000 in the reliable source today, the money is instead hidden in a mattress. Three years later, the $100,000 is withdrawn from the mattress. The money has not produced any interest in the 3-year interim – it has retained only its nominal value of $100.

The NPV of this approach is given by:

$$NPV = -100,000 + 100,000/(1+0.1)^3 = -24,870$$

Thus, a present value of $24,870 would be lost by stuffing the cash in a mattress rather than investing in the reliable source. This is the opportunity cost of "investing" in the mattress approach rather than in the reliable source.

Seen another way, the mattress "investor" could have been equally well off financially if she had spent $24.87 for some other purpose at year 0; and deposited the balance of the money ($100-$24.87=$75.13) in the reliable source rather than the mattress.

Come year 3, this investor could withdraw $100 from the bank – exactly the same amount that they would have been able to withdraw from the mattress.

By making this investment in the bank rather than the mattress, the investor enjoys the use of spending almost a quarter of the quarter of the total money that would have been invested in the mattress up front, while recouping an identical amount of money in year 3. Investing in the mattress is thus sacrificing the use of the $24.87 – hence the negative NPV.

The discussion and example above assumed annually compounding discount rate. However, more frequent compounding periods can be incorporated into NPV analysis by choosing a suitable discount rate. If I is the annual interest rate and there are n compounding periods each year, the effective interest rate for the entire year is given by:

$$\text{effective annual interest rate} = (1 + i/n)^n \tag{7.6}$$

Note that in (7.6) as n approaches infinity (i.e., continuous compounding similar to a savings account), the effective rate approaches e^i. Over t years, the result is e^{it}.

Following is a list of the main issues and practical rules for evaluating a project using the NPV method.

7.2.2 Choice of Discount Rate

The choice of a discount rate is very important in NPV analysis.

When the interest rate of a reliable source is used to discount the project expected cash flow, the NPV denotes the value of the project compared to a reliable investment. Similarly, if the discount rate is referred to as a minimum attractive rate of return (MARR), a positive NPV is an indication that the project is to be undertaken: the higher the NPV, the greater the attractiveness of the project.

The minimum attractive rate of return (MARR) is usually the minimum acceptable discount rate the investor is willing to accept for the risks associated with a given project. In general, the MARR can be represented as:

$$MARR = r_f + r_i + r_r \tag{7.7}$$

where r_f is the "risk-free" interest rate offered by a reliable source such as a government bond or a similar security, r_i is the inflation rate, and r_r encompasses market risk, industry risk, firm specific risk, and project risk.

Thus, the minimum attractive discount rate for a given project may or may not be appropriate to use for another project. If two projects are very similar in capital structure and risk, then it may be appropriate to use the same discount rate. However, it is often the case that projects are of varying levels of risk and this must be reflected in the choice of discount rate; the discount rate is the primary means of capturing the risk associated with a project.

If the leverage ratio of a project is fairly constant over time, the weighted average cost of capital (WACC) is often a good approximation of the appropriate discount rate, but care must also be given to not blindly apply the WACC. As shown in the paragraph above, derivation of the WACC attempts to reflect the cost of equity, the cost of debt, and the target leverage ratio of a given project.

7.2.3 IRR Versus NPV

An often-used evaluation method similar to NPV is the Internal Rate of Return (IRR).

The simplest definition of IRR is the discount rate required to achieve an NPV of zero for a given stream of cash flows.

Example – Computing the IRR

Consider the following discounted cash flow stream:

$$NPV = -20 + 10/(1 + i) + 20/(1 + i)^2.$$

The IRR for this cash flow stream is found by setting the NPV equal to zero and solving for $i = IRR$ as follows:

$$0 = NPV = -20 + 10/(1 + i) + 20/(1 + i)^2$$
$$i = 28\% = IRR$$

While the value of i can be determined through a formula (the quadratic formula) for this particular case, typical cash flows would in general require solution via numerical techniques (e.g. specific calculation using an electronic spreadsheet or via hand-driven iteration through values of i)

The evaluation methodology using IRR then becomes "accept a project with an IRR larger than the MARR" or "maximize IRR across mutually exclusive projects" (Hendrickson 2008). The concept is similar to NPV analysis in that the project has an IRR below the MARR, the project would have a negative NPV.

It is common for the IRR and NPV approaches to produce the same ranking of projects. However, IRR ignores the capacity to reinvest and captures a project's rate of gain, not the size of gain (Brealey et al., 2006). Thus a more appropriate method is to use IRR and NPV complementarily instead of independently. Given a list of many projects, requiring an IRR greater than the MARR hurdle can be used to give an idea of which projects should be further explored. NPV analysis can then be used to further narrow the list by choosing projects in descending order of NPV.

It is important to remember that IRR is defined in terms of NPV, and that NPV captures everything the IRR method does, and more. Thus, if a project cash flow has the greatest NPV but not necessarily the highest IRR, it is opportune to verify the equity capital net cash flow profitability. As shown in the example below, once the debt leverage has been determined for a project, the equity cash flow includes cash outflows related to the equity investment into the project and the later net income cash inflows: then the higher Equity NPV project should be chosen.

Example – Equity profitability

Suppose a $1 million investment can generate A or B cash streams. Both have same cost of equity (i), equal to 15%, and same cost of debt at 10%.
The projects result in the following NPV and IRR for both project cash flow and equity cash flow.
Both projects have same financial structure covering the $1 million investment with 30% equity and 70% debt, plus interest charges. Accordingly, the resulting WACC is the same.

	A	B
NPV(i)	154.678,74	105.463,86
IRR	16,4%	16,6%
NPV (wacc)	154.678,74	105.463,86
Equity NPV (i)	285.707,21	367.947,29
Equity IRR	35,1%	54,5%

The comparison, in the table, suggests that project A is better than B in terms of project NPV (and not in terms of IRR) but, if we consider the equity perspective B is much better than A, when judged in terms of both Equity NPV and in terms of Equity IRR.

Results reflect a quicker return on equity for B.

Also, while the NPV of a project is unique and well-defined, the IRR of a project can be ambiguous. This can occur, for example, when a project exhibits significant alternating periods of high future expenses and revenues.

Methods other than IRR and NPV are sometimes used to quantitatively evaluate projects.

The minimal length of time for a project's benefits to repay its costs, or the project's payback period, is sometimes used as a secondary assessment. This method and its discounted version, the capital recovery period, ignore the costs or benefits occurring after the payback period.

The equity payback period is the point in time at which the nominal cumulative cash flows equal the equity of the project. This method also ignores the events occurring after the equity payback period. These methods are typically used in an informative manner rather than as a comparison and decision tool.

Discounted cash flow analysis assumes some certainty regarding future cash flows. The discount rate attempts to capture any uncertainty stemming from the risk of a given project, but the valuation produced is only as good as the estimates and projections provided.

Furthermore, DCF analysis considers only quantifiable monetary benefits. The social benefits occurring from the construction of schools or hospitals, or the strategic benefit of a long-term partnership or entry into new markets are further examples of benefits DCF fails to capture properly. In such cases, benefits/cost ratios can be used to assess a project feasibility and accept if the ratio is greater than one (benefits > costs). Problems still persist because it can be difficult to determine whether something counts as a "benefit" or a "negative cost": to overcome the challenge it is opportune to take into account accrued costs assuming that project was not built (MIT Open Courseware). Alternatively, cost-effectiveness models look at non-economic results of the investment such as $/life saved or $/Case of illness averted.

References and Additional Resources About Feasibility

Brealey RA, Myers SC, Allen F (2006) Principles of corporate finance. McGraw-Hill, New York, NY

Construction Specifications Institute http://www.csinet.org/

Halpin DW, Woodhead RW (1980) Construction management. Wiley, New York, NY, Chapter 8

Finnerty JD (2007) Project financing asset-based financinal engineering. Wiley

Hendrickson C (2008) Project management for construction, Chapter 7. http://www.ce.cmu.edu/pmbook/07_Financing_of_Constructed_Facilities.html

MIT Open Courseware http://ocw.mit.edu/NR/rdonlyres/Civil-and-Environmental-Engineering/1-040Spring-2004/ABF26C4A-8572-498D-ACE3-98D2E8AD0685/0/l3prj_eval_fina2.pdf (05.17.08)

Sullivan WG, Bontadelli JA, Wicks EM (2001) Engineering economy. Prentice Hall, Upper Saddle River, NJ

Chapter 8
Planning and Scheduling

8.1 Project Planning: Breakdown Structuring

Construction project planning is a method of determining "What" is going to be done, "How" things are going to be done, "Who" will be doing activities and "How much" activities will cost.

In this sense planning does not cover scheduling, which addresses the "When", but once planning is complete scheduling can be done (Fig. 8.1).

Fig. 8.1 General framework for the planning process

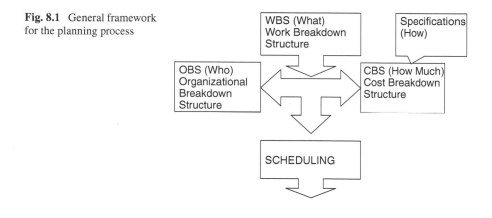

8.1.1 Work Breakdown Structure – "What"

When projects are simple, consisting of few defined activities, it might be possible for a single person to grasp the total construction effort with little difficulty. Unfortunately, most projects for which formal plans are prepared tend to be defined with dozens or even hundreds or thousands of activities: the larger the project, the greater the number of activities and higher the level of detail managers have to handle.

A. De Marco, *Project Management for Facility Constructions*,
DOI 10.1007/978-3-642-17092-8_8, © Springer-Verlag Berlin Heidelberg 2011

When a project plan consists of numerous activities, it is often advisable to organize the activities in some way to allow communication of plan information to others and to maintain an understanding of the various aspects of the project. While there are many ways that a plan can be organized, one common practice is the Work Breakdown Structure (WBS).

The WBS is a convenient method for decomposing the project complexity in a rational manner into work packages and elementary activities. Some firms prefer to use a standard means of identifying work packages common to all similar projects. These work packages are then coded so that both costs and the schedule can be controlled. A common numerical accounting system is then applied to the activities, so that the coding indicates factors such as the type of material involved or the physical location within the project.

In essence, the WBS divides and subdivides a project into different components, whether by area, phase, function, or other considerations. The highest level in the WBS consists of a single element, the project. At the next level, there may be only a few elements or items. Naturally, the further one goes down within the WBS, the greater the granularity of decomposition and the amount of detail. Regardless of the means used to define the elements, individual tasks are to be defined for the lowest level in the hierarchy or at the greatest level of detail that is required to adequately manage and control the construction process. The level of detail used will be determined by the scheduling needs and the roles of the people viewing the WBS. For example, if one is a homeowner and having a house built, one is interested in the completion date of the project, but a subcontractor will be primarily interested in information related specifically to the task this has direct responsibility.

Commonly there are three main types of WBS, namely, the Project WBS, Standard WBS and Contract WBS.

The Project WBS is an operational tool usually prepared by contractors to monitor and control the work (Fig. 8.2 is an example for a new stadium construction project).

A standard WBS is a breakdown structure of activities carried out in the past for a similar project: the past project WBS can that can be used as a template for the new one. Figure 8.3 shows the highest levels of a sample template WBS that might be used for turn-key construction of an ordinary industrial building.

A contract WBS is agreed between owner and contractor. This is a decomposition of the scope of work into the main elements that will be used for progress measurement, control and payment of the contract price. It may include less detail than a Project WBS.

To summarize, WBS is a deliverable-oriented decomposition of the project scope (Project Management Institute 2008) until a sufficient level of granularity enables easy definition of all information required to execute and manage detailed tasks.

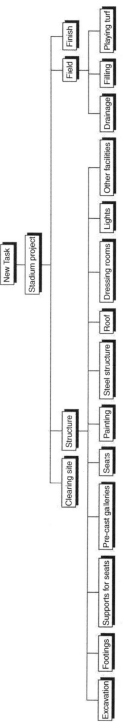

Fig. 8.2 Example of Project WBS of a new stadium construction project. WBS chart based on the case story by E. Turban and J.R. Meredith "The Sharon Construction Corporation" from Meredith and Mantel (2006)

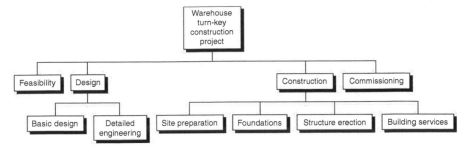

Fig. 8.3 Example of standard WBS for a warehouse construction project

8.1.2 Organizational Breakdown Structure – "Who"

Once what needs to be done is defined, it is necessary that all human resources required to perform the project are identified. Depending on the portions of work scope, the project may need engineering skills, procurement capabilities, construction labor, management staff, etc.

The Organization Breakdown Structure is a practical method to decompose the pool of human resources needed to execute all of the tasks into different competence areas and then into project roles, independently of the number of individuals that will be assigned the specified role (Fig. 8.4).

The OBS is prepared with the idea that each task in the WBS must be assigned to a role or committee of roles. In other words, roles are allocated to detailed tasks

Project resources
- o   Internal resources
  - ▪   Project management team
    - •   Project Manager
    - •   Site Manager
    - •   Scheduler
    - •   Site inspector
  - ▪   Engineering
    - •   Project Engineer
    - •   Architect
    - •   Civil Engineer
  - ▪   Construction labor
    - •   Foreman
    - •   General worker
- o   External resources
  - ▪   Subcontractors
    - •   Electrical
    - •   Plumbing
    - •   Etc.
  - ▪   Material suppliers
    - •   Steel pre-casted structures
    - •   Construction commodities

Fig. 8.4 Example of OBS

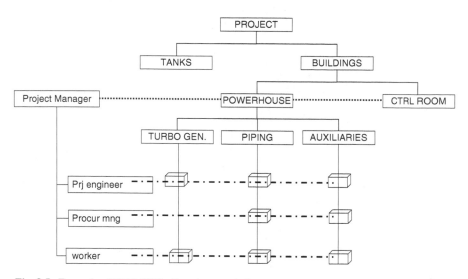

Fig. 8.5 Example of WBS/OBS allocation matrix for a power station construction project

with a specified number of resources and related estimated work load required to perform the task.

Figure 8.5 is a simplified depiction of how a WBS and OBS correlate. It is imperative during the monitoring and controlling phase of the construction project that there is a responsible person for each activity.

To summarize the planning process thus far, we have created the WBS and have now incorporated a responsible committee or person to each element of the WBS. In essence, from activities developed in the WBS we have allocated resources through the OBS.

8.1.3 Cost Breakdown Structure – "How Much"

Now that we have discussed "what" is going to be accomplished through the WBS and "who" is going to perform activities through the OBS, owners and contractors want to know how much things will cost.

Determining the cost is done through the Cost Breakdown Structure (CBS). The CBS is a system for dividing a project into hardware elements and sub elements, functions and sub functions and cost categories. It is a hierarchical structure that classifies resources into cost accounts, typically labor, materials, and other direct costs. In addition it represents the economic breakdown of the project into budgets per work package. This will allow the project manager to track project progress and expenditure according to planning breakdown of activities and responsibilities.

A CBS includes all direct full cost of labor, material, as well as the so-called project overhead, which is still a direct cost required to execute the project. Project overhead embraces the cost of construction equipment (usually under the terms of average amortization of construction assets), project management, design services, permits and insurance fees. CBS does not have to include the company's overhead not associated with the project, such as general office salaries, utilities, insurance, taxes, interest, and other expenses out of the direct control of the project team, but rather inherent with corporate top management's action.

There are two main approaches to direct cost breakdown structuring. Which is used in a particular circumstance depends on the different purposes of cost accounting.

The first one makes use of the WBS as the project cost control structure, so that the CBS and WBS are the same structure and each cost account is consistent with a work package or detailed task. In other words, the accounting structure is the same WBS that has been filled with cost information: the end result is a hierarchical structure of cost to be used by the project team for both budgeting, accounting and control. With this kind of CBS, Activity Based Costing (ABC) method drives both estimation of budget and accounting of actual expenditures. The advantage is that project budgeting and tracking develop on the WBS exactly in the way the facility is going to be built, with detailed analysis at the final level of decomposition of the WBS: the cost of an elementary activity may include a combined summation of full cost of labor, quantity of material, equipment, and lump-sum cost of subcontract or service.

To define the budget, a different methodology may apply to parts of the breakdown depending on the specific nature of items or elements. Subcontractor quotes are of practical use when a specialized subcontractor is assigned a job. Quantity takeoffs are obtained by multiplying the measured quantities by the unit cost, which includes material, equipment and labor as a whole. Challenges here are the tremendous detail complexity of line items, the dependence of the estimated quantities on construction methods, and the determination of unit cost based on historical data. Material takeoff estimation is needed when data about unit costs for complete installation of materials are unknown. For each line item in the cost breakdown, a quantity of material required, Q, must be determined. For each item the unit cost of material, M, can be estimated using quotes from local material suppliers. For most line items equipment is involved in the construction process, and an equipment rate of cost, E_M (cost per unit of material), must be determined. In addition, labor costs – which are often greater than material cost – must be incorporated by multiplying the hourly wage rate, W, and the labor cost per unit of material (productivity) L. Combining these factors in the following equation produces an estimate of the direct cost for a given item:

$$\text{Total cost \$} = Q * (M + E_M + W * L) \tag{8.1}$$

Regardless of the method applied, careful consideration of wages and productivity has to be taken into account for appropriate detailed budgeting. Labor cost estimation W) is affected by several components, namely wages, insurance, social security, benefits and premiums. Productivity (L) impacts a project in many ways.

At the beginning of a job workers will typically have lower productivity on account of inexperienced with the particular routine to be followed. As time progresses they become more efficient in their work with repetition due to the effects of learning: an effect expressed in learning curves (Kerzner 2001).

However, some projects have little repetitive tasks, and therefore must account for this factor in the project estimate. When productivity is less than initially expected a project may begin to fall behind schedule. As a result the project manager may increase pressure in order to finish more quickly. However, as hours per day of work increase, worker productivity per hour is known to decrease. Productivity also suffers greatly over the medium- and long-term as workers become fatigued and lose motivation. This reciprocal process can be damaging to the success of a project if it is not realized. Productivity can be measured, but the results of corrective actions are highly uncertain. In this realm, a project manager with good experience and a good understanding of his personnel can identify problems and attempt to remedy them – ideally before the time such problems begin to be evident in project reports and failure to meet the schedule of values. Lost time due to low productivity can be incorporated into an updated cost estimation, but prior to construction this additional cost is most easily calculated as a contingency. Applying probabilistic models to estimation calculations allows planners to gain a deeper insight into the effects of uncertainty in costs.

Even if the probabilistic distribution is not fully known, the effects of changing the range of outcomes can help planners see where major problems may occur. Finding the variance of just one portion of a project can give insight into the effects of increased costs will have on the total project cost (see Section 8.5 "Uncertainty").

This practical first way of accounting for cost based on project activities is usually adopted when a firm does not have a specific cost control accounting system.

A second approach to CBS budgeting is to use the corporate multiple-project cost control structure as the project cost accounting system. With this method, each WBS activity has to be associated with a cost account by the means of a cost code.

The coding system may be a firm-specific or a common standardized one, such as the Master Format developed by the Construction Specifications Institute of the United States, the ISO UniClass, the German KKS valuable for power plant construction, or the Construction and the Engineering Information Classification System.

An illustration of how a cost code is often represented is below in Fig. 8.6. The cost code reflects the WBS decomposition and contains several subfields: the first is the project code for the first level of the WBS, the second code physically identifies areas or sub-facilities, then the Masterformat code describes the activity, and the final digit represents the distribution code (0 = Total, 1 = Labor, 2 = Material, 3 = Equipment, 4 = Subcontract).

CBS is also utilized in different approaches by means of delivery.

In case of a Design Bid Build delivery system, the Contract WBS is the same as the Contract CBS because schedules of values are paid unit price by performed units. As a result, most often the contractor's own CBS used for cost accounting is quite different from the contract CBS; the project operating WBS will also differ

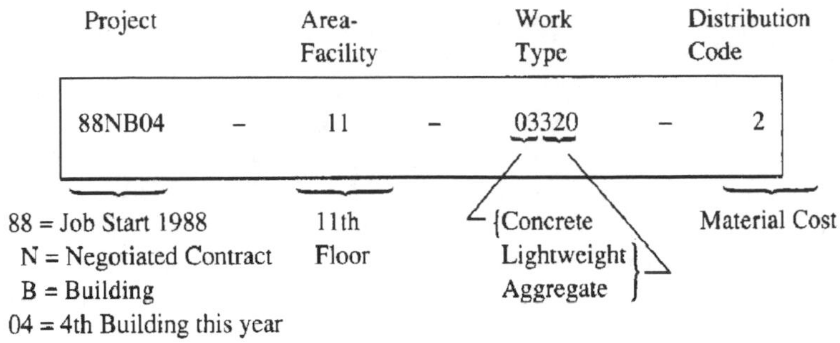

Project	Area- Facility	Work Type	Distribution Code

```
┌─────────────────────────────────────────────────────────────┐
│   88NB04      –      11      –      03320      –      2       │
└─────────────────────────────────────────────────────────────┘
```

88 = Job Start 1988   11th   ⌐{Concrete   Material Cost
N = Negotiated Contract Floor   Lightweight
B = Building   Aggregate ⌡
04 = 4th Building this year

Fig. 8.6 Example of cost code integrating the UCI/CSI MASTERFORMAT

from the Contract WBS. In such circumstances, the solution is to keep the revenue and cost separate.

Instead, in a Design-Build or Turnkey project, the Contract WBS is prepared by the contractor himself and therefore it is equal to the Project WBS. The sum of the contract work packages is paid cost plus and the contract price is paid on a project progress basis. Since the revenue is a function of cost, then the project WBS should reflect the CBS, if corporate cost control is required. If this is required, then it is recommended to use the higher level of CBS codes, and then break down according to the job needs.

In summary, planning tasks include scope of work definition and budgeting, as a fundamental precursor to scheduling the estimated time to perform a project, as discussed in the following paragraphs.

8.2 Deterministic Scheduling Principles

Deterministic scheduling is just one of the many tools available to project managers during the planning stages of a project. However, it may be one of the most important because it both lowers chance of delay and assists in recovering from delay, resolving responsibility. Indeed, delays often result simply from poor planning.

Accurate scheduling assists in reasoning about a huge number of details (e.g. thousands of activities), and determines a lot of things, including expenditure estimates for crews and materials, expected opening dates (there may be situations where a strict opening date is highly important, such as a new production facility), scheduling changes with sufficient flexibility to not affect the completion date, and others.

Scheduling also allows for accountability. Setting milestones from the beginning allows for the project managers or the owners to pinpoint exactly what went wrong and who or what was responsible for a delay.

A schedule is also a good communication tool, between the managers, the owners, investors, and the general public. Schedules give an overall sense of the project's expected progress. Without schedules, it's much more difficult to explain to someone unfamiliar with the project what is expected to take place.

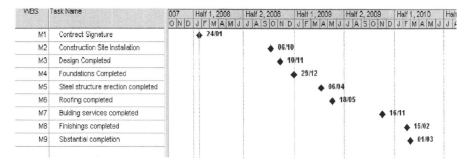

WBS	Task Name	007	Half 1, 2008	Half 2, 2008	Half 1, 2009	Half 2, 2009	Half 1, 2010	Half
		O N D	J F M A M J	J A S O N D	J F M A M J	J A S O N D	J F M A M J	J A
M1	Contract Signature	◆ 24/01						
M2	Construction Site Installation		◆ 06/10					
M3	Design Completed		◆ 10/11					
M4	Foundations Completed		◆ 29/12					
M5	Steel structure erection completed			◆ 06/04				
M6	Roofing completed			◆ 18/05				
M7	Building services completed				◆ 16/11			
M8	Finishings completed					◆ 15/02		
M9	Sbstantial completion					◆ 01/03		

Fig. 8.7 Example of master schedule under the form of a milestone chart

A schedule can also be used a contractual tool. Some payment schemes are based on scheduling. Some offer incentives for finishing the job on time or ahead of schedule. With an accurate schedule, these sorts of incentives can be offered fairly in the contract from the very beginning. Also, in the case of a lawsuit, a good schedule can serve as great evidence in support of the parties.

To put a schedule into effect it is recommended to avoid any imbalanced use (such as to use it early on and discarding later), to game for liability reasons (i.e.: schedule as a biased document to support the originator's rights), or to use for central PM office only. In contrast, schedules should be used as shared management tools to get to an integrated point of view for both the owner and the contractor.

Schedule documents can be subsumed mainly in two types. One is the *Master Schedule* that is used as the contract baseline, usually under the form of a milestone chart, as in Fig. 8.7.

The other is the *Project Schedule* which is used to monitor and control the actual progress of the project. This schedule is usually based on the WBS and is very meticulous. It usually includes detailed plans, such as engineering schedules, construction sequencing, quality-assurance activities, as well as procurement plans.

For example, a procurement detailed schedule involves trying to schedule when materials will be ready and available on site for installation. This is often difficult to estimate, especially for custom built items, though it is very important to keep work on pace. Without the proper materials on site, workers may be sitting around and money will be spent on entertaining them.

For the project schedule, typically there are revisions performed on a weekly, monthly, or other periodic system. Then, these revisions are used to track progress against the original schedule. This allows for the managers to make any changes, if necessary, to the work (see later Chap. 9).

8.3 Scheduling Systems

So how do we schedule? There are several forms of schedules and several methods used to determine accurately the schedule. The following methods will be discussed in greater detail in the following: task matrix, Gantt chart, network diagram, and line-of-balance scheduling.

8.3.1 Matrix Scheduling

Matrix scheduling is fairly simple. It is usually used for small, less complex projects because of this simplicity. It also doesn't have a clear way of showing interactions between different tasks. Table 8.1 shows an example of matrix scheduling.

Table 8.1 Example of matrix scheduling

Task	Original schedule	Review week 8	Review week 16	Actual week 22
Mechanical design – Start	1	1	1	1
Mechanical design – Finish	10	10	10	10
Electrical design – Start	8	8	8	8
Electrical design – Finish	16	16	18 (delay 2)	18
Software dev – Start	14	14	16 (delay 2)	18 (delay 4)
etc.				

This example shows an original schedule and then makes comparisons based on the reviews every 8 weeks. In the week 16 review, the delays began in the electrical design. However, it is unclear whether the design review caused the delay in the SW development, or whether that was due to something else. The next method we discuss shows more clearly those relationships.

8.3.2 Gantt Chart Scheduling

Figure 8.8 shows a basic Gantt chart. Here we begin to see a more clear relationship between tasks, though not completely. For example, we know that design has to take place before construction, but construction could begin before the design is completed. So there is some intuition as to which tasks are related, but not an explicit statement of dependencies. Each bar represents the amount of time that its respective task will take.

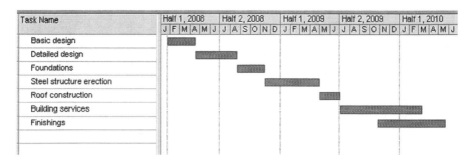

Fig. 8.8 Example of a Gantt chart

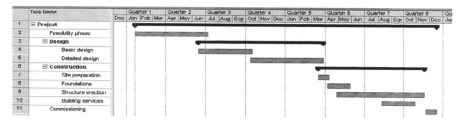

Fig. 8.9 Gantt chart schedule of Fig. 8.2 WBS

This form of scheduling is far superior to that of the matrix scheduling in that it's more effective as a communication tool. This type of chart is very easy for anyone to understand and allows for the owner or manager to more effectively communicate how the project will proceed.

There may also be WBS levels of scheduling. Figure 8.9 illustrates that idea.

However, we need a more detailed way of showing relationships of activities.

8.3.3 Network Diagramming

This method is a most robust way of showing and calculating a schedule. Using this method of scheduling, it is fairly easy to use software tools to calculate project duration and optimize allocation of labor and resources. It is also relatively easy to find the areas in the schedule which are more flexible to change.

Basically, the process of constructing a network system is composed of the following stages:

- First, the tasks are drawn from WBS work packages and assigned expected deterministic duration, estimate cost, and resources as discussed in Section 8.1. The method for obtaining the deterministic durations may vary depending on the task, but mostly it's a factor of amount of work to be performed, productivity, number of resources and equipment used. Costs can also be assigned to each task based on the original cost estimates or trough assignment of human resources, materials and equipment to each task. In any case, the common assumption in deterministic estimation is that all activity attributes can be determined as certain values with very little margin of error (in a later section we will discuss about probabilistic estimation of task attributes).
- Second, each task is assigned precedence relationships with other tasks. In other words, if task B cannot be started until task A is finished, that relationship is defined in this method.
- Then, the network diagram is solved and optimized using various ways, such as Critical Path Method, Precedence Diagramming Method and Program Evaluation Review Technique. This often implies iteration: if the solution of the network acceptable in terms of total project duration and resource allocation, then terminate. If it is not acceptable, it is needed to impose dependencies or added/reduced resources.

Table 8.2 Task list from
project breakdown

Task name	ID
Feasibility study	1
Basic design	2
Detailed design	3
Site preparation	4
Foundations	5
Structure erection	6
Building services	7
Commissioning	8

Let us first discuss the precedence relationship process. The first step is to list the activities that need to be performed. This is done by taking the tasks defined in the WBS and listing them. The following shows an illustration of listing the activities from WBS detailed in Table 8.2.

Once the tasks are listed, one has to assign precedence relationships. Sometimes activities can overlap; sometimes they have to occur in series. So we define a matrix of precedence to capture this idea (Table 8.3).

As per the above diagram, the relationships between activities reflect the constraints in sequencing the tasks, such as regulatory or contractual, physical or functional, financial, managerial, and environmental constraints. Also, resource availability may restrain multiple tasks in parallel: for example, if only one crew is available to perform the job all construction tasks have to be performed in series.

Finally, representation is required to capture the above relationship matrix in diagram form, again to allow the scheduler to clearly understand how activities will unfold. There are two ways of graphical diagramming: Activities on Arrows (AOA) or Activities on Nodes (AON).

AOA representation keeps similarities to a Gantt format. In this method, Nodes represent start and finish events for each activity. Arrows represent the tasks that need to be done to get to the next activity. The diagram in Fig. 8.10 shows a sample task depicted with AOA mode.

A problem that sometimes arises using this method is that we need to create "dummy nodes". These nodes arise when one task has two or more precedent activities, as in the example. Because a node may only have one incoming arrow, dummy nodes need to be created. Figure 8.11 illustrates that idea.

Once all the nodes are accurately represented, one may construct the final diagram. Figure 8.12 shows an AOA representation of the previous schedule.

AON is the method most popularly used in today's project planning software programs, such as Microsoft Project or Primavera. A task is represented in Fig. 8.13.

Fig. 8.10 Activities on
arrows representation (AOA)

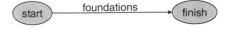

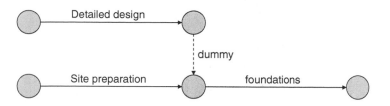

Fig. 8.11 Dummy activities are necessary for AOA representation

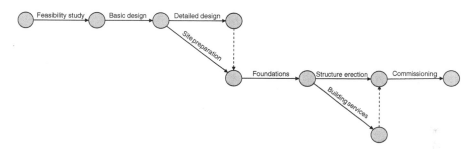

Fig. 8.12 Example of a network diagram with AOA representation

Thus, from the example in Fig. 8.13, the resulting AON diagram is graphed in Fig. 8.14.

Also, as with the Gantt chart, we are able to illustrate a hierarchy of networks, as in Fig. 8.15. So setting different levels of hierarchy may help in presentation, where a client or a top manager may not need to know the details of the construction, but may just want an overall view of the process.

AON representation is also closely related to the so-called "Precedence Diagram Method" (PDM) or "Bubble Diagram Method" that allow for representing richer

Fig. 8.13 Activities on nodes representation

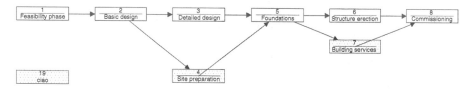

Fig. 8.14 Precedence diagram for concrete footing construction

Table 8.3 Matrix of precedence relationships between tasks

Task name	ID	Feasibility study	Basic design	Detailed design	Site preparation	Foundations	Structure erection	Building services	Commissioning
		1	2	3	4	5	6	7	8
Feasibility study	1								
Basic design	2	X							
Detailed design	3		X						
Site preparation	4		X						
Foundations	5			X	X				
Structure erection	6					X			
Building services	7					X			
Commissioning	8						X	X	

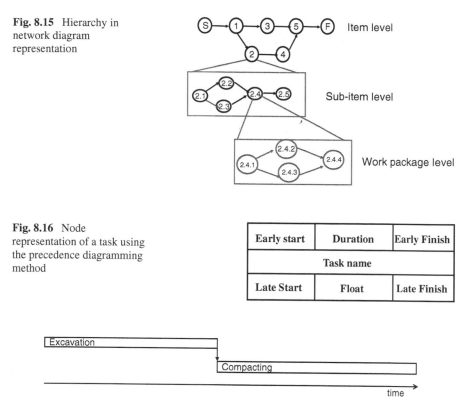

Fig. 8.15 Hierarchy in network diagram representation

Item level

Sub-item level

Work package level

Fig. 8.16 Node representation of a task using the precedence diagramming method

Early start	Duration	Early Finish
Task name		
Late Start	Float	Late Finish

Excavation

Compacting

time

Fig. 8.17 Node bar chart with added graphical precedences

semantics, such as early/late start and finish events of activities (Fig. 8.16) and varied possibilities for setting diverse constraints between tasks (start-to-start, finish-to-finish, start-to-finish, finish-to-start). PDM nuances will be better discussed in the following paragraph coping with critical paths and time floats.

Another way of modeling network dependencies is using bar charts with precedence notation, as shown in Fig. 8.17. By adding arrows to a Gantt chart, it is possible to capture the AON precedence relationships, while being able to maintain the easy-reading of the Gantt chart.

8.3.4 Line-of-Balance Scheduling

Finally, another way for graphical representation of scheduling is the Line-of-Balance (LOB) method otherwise called *Chemin-de-Fer* from the French national railroad company (SNCF Societé Nationale des Chemins de Fer) who widely uses this technique to schedule linear works such as railroad tracks, roads, and tunnels.

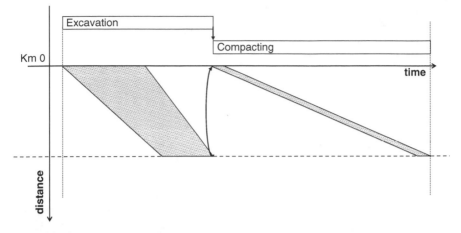

Fig. 8.18 Line-of-balance representation of scheduling

In a LOB graph, time is usually plotted on the horizontal axis and space on the vertical one. This diagram allows for representing the production rate of an activity: the slope of the production line is expressed in terms of units of distance per time (i.e. km/day).

From the example in Fig. 8.18, it is also clear that the production rate for each kilometer of excavation is variable depending on several parameters: as distance increases, time to perform excavation decreases. This may depend on several factors such as use of more resources, decrease in volumes of excavation (the dig may be less deep or narrower), or/and more efficient technologies.

8.4 Critical Path Method

There are different scheduling practices depending on whether the duration of activities is considered to be deterministic or probabilistic. Under the deterministic assumption, the most used is the Critical Path Method (CPM) and its strictly derived Precedence Diagramming Method (PDM).

The CPM consists of specifying the activities to be carried out and its associated information (such as duration) and running a scheduling algorithm in order to yield some scheduling recommendations and constraints.

The CPM runs on a network-based scheduling system. The basic steps to follow are: define activities from WBS work packages, estimate the cost, duration and resources for each one of the activities and define the precedence relationships between them. Once all is clearly defined, the system needs to be iterated in order to optimize and manage the network, using the CPM algorithm. If the results obtained are acceptable, the iteration must stop. Otherwise, some extra dependencies need to be added or some additional resources need to be considered.

The CPM algorithm runs either on AOA diagrams or on AON diagrams and it computes Early and Late Finish as well as Early and Late Start for each node. Late

Table 8.4 Precedence table of a roof construction project

#	Task name	Duration (weeks)	Predecessor
1	Site preparation	2	
2	Excavation and Foundations	7	1
3	Structures	2	2
4	Roofing	2	3
5	Enclosures	4	3
6	Building services	5	5
7	Finishing	6	4; 6

Start and Late Finish for each activity is defined as those latest dates to start or complete an activity without delaying the project duration as a whole.

For each activity, the difference between the Late Start and the Early Start (as well as between Late Finish and Early Finish) constitutes the so-called "Float".

The CPM algorithm consists of two phases or passes:

- Forward pass determines Early Start and Finish of activities. Because all preceding activities must finish before a successor, early start of a given node is the maximum of early finishes of preceding nodes. As a practical example, the forward pass determines the shortest time to complete a sequence of tasks.
- Backward pass determines Late Start and Finish dates. Because preceding activity must finish before any following activity, late finish of a given activity is minimum of late starts of successors. In practice, given the final completion time of a sequence of tasks, the backward pass allows calculating the latest point in time the sequence has to be initiated.

Both notions are quite common-sense reasoning that we use all the time for daily life tasks (e.g. we use the forward pass to figure out what is the earliest time we could meet someone, or use the backward pass to know at what time we need to leave for making an airplane on time).

Below is an example for the construction of a small residential unit. Consider the project described with the precedence matrix in Table 8.4.

With these tasks and their predecessors in mind, the network diagram looks like the one in Fig. 8.19:

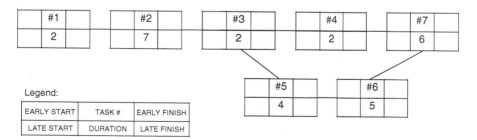

Fig. 8.19 Network diagram of the roof construction project presented in Table 8.3

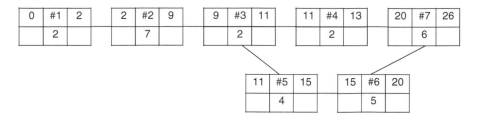

Fig. 8.20 Forward pass allows for calculating early start and finish dates of activities

Some conclusions can be extracted from application of the forward pass principle in Fig. 8.20: the Early Finish date of the project is 26 weeks.

The next Fig. 8.21 shows the second phase of the CPM algorithm: the backward pass. Now, as we know the durations of the activities, we subtract them from the Late Finish to get their Late Starts.

With also Late Start and Finish dates in hand, it is possible to calculate floats for each one of the activities. In the project above, for example, activity #7 has no float, while activity #6 has a 7-week float.

After all the network is solved, we just need to look at that path whose activities have no float. This path is defined as the *Critical Path* (CP) and it is the longest of all paths in the network system. In the example above, the CP is the one comprised of activities #1-2-3-5-6-7.

In all projects where the total finish date is calculated as the late duration of the network, there is at least one critical path, and the activities in this path must be completed on time, otherwise the entire project will be delayed.

Sometimes, projects have a later contract deadline than the one obtained from solving the network. In such fortunate circumstances, there is no critical path in a strict sense. Yet, it is opportune that a new project timeline is set to be finished with the longest path of activities, so that a time buffer, from timeline completion to contract deadline, is available as a contingency.

The CP determines the minimum time required to execute a project. However, two aspects of this algorithm need to be considered: first, we have to pay special attention to near-critical paths (those paths with low floats), and second, the critical path evolves over time as activity actual durations unfold. Finally, since there is no

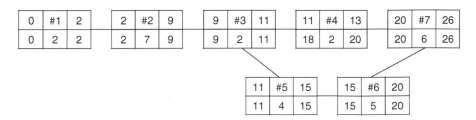

Fig. 8.21 Backward pass allows for calculating late start and finish dates of activities

float in the critical path, there is no flexibility and, thus, some contingency buffer should be planned ahead.

Therefore, the notion of float assumes great importance. Intuitively, the float measures the leeway in scheduling: it is somewhat a degree of freedom in timing for performing a task.

There are two different types of float:

1. the *Total Float* of a path, represents the maximum amount of time that will not delay the overall project;
2. the *Free Float*, for each activity, represents the amount of time an activity can be delayed without delaying the start of its successors. Closely similar is the Independent Float, which is defined as the Free Float in the worst-case finish of all its predecessors.

In light of this definition, a critical path is that with a total float equal to 0. Those paths with a total float greater than 0 are called sub-critical and those with a float less than 0 are called hyper-critical. In this latter case, it is necessary, either by increasing the number of resources and the productivity rate or by changing the equipment and the technology, to expedite the network and bring the hyper-critical paths to critical, at least.

One way to rank all the paths in order to know which ones need more attention is by using the priority index, defined as:

$$\lambda = \frac{\alpha_2 - \beta}{\alpha_2 - \alpha_1}(100\%)$$

where α_1 is the minimum total float, α_2 is the maximum total float and β is the float of the considered path. In this way, we can classify all paths and pay attention as λ is high.

Consider this Example. A Project has 4 Paths with the Following Total Floats:

Path 1: $b_1 = 0$ days, which is equal to minimum total float α_1
Path 2: $b_2 = 10$ days, which is equal to maximum total float α_2
Path 3: $b_3 = 5$ days
Path 4: $b_4 = 2$ days

The priority indexes for the four paths will be:

$\lambda_1 = (10–0)/(10–0) = 100\%$, Critical Path
$\lambda_2 = (10–10)/(10–0) = 0\%$, the less critical path of the project
$\lambda_3 = (10–5)/(10–0) = 50\%$, medium critical
$\lambda_4 = (10–2)/(10–0) = 80\%$, near-critical

8.4.1 Float Ownership

Tensions and disputes often occur between owners and contractors over the "own-ership" of the float. The problem arises when, on the one hand, owners seek to push contractors on a tight (and sometimes unrealistic) schedule, while, on the other hand, contractors seek flexibility in their projects.

Thus, the owner seeks lower risks by getting the work done the earliest (because too many late starts may jeopardize the overall project duration) and, in this endeavor, the owner may impose unrealistic short schedule to the contractor. The owner may also use the contract to limit the flexibility of the contractor by specify-ing the owner rights to use the float, to select the scheduling procedures or to object to unreasonable durations.

On the other side, the contractor will try to artificially create a schedule with many critical and near-critical paths by deliberately inflating durations (so that they can charge extra money if the owner requires them to speed up) or by inserting artificial precedence constraints (so that the contractor can charge an extra amount of money if the owner requires them to change "the way of doing things").

Fisk (2003, pp. 362–364) proposes a proper distribution of floats that may help in solving tensions and better understand who is responsible for delays, as follows.

There are two common ways of distributing the available float all over the non-critical activities: straight-linear and distributed. To present those methods, consider the example in Fig. 8.22.

The critical path (marked grey in the figure) has duration of 30 days. The black-marked path has a total duration of 24 days and it is comprised of activities #1 (duration 3 days) and #2 (21 days). Therefore, the total float for the black path is 6 days (30–24 days).

A first way to distribute the total float is by using a straight-line method. That is, distribute the float proportionally to the duration of each activity of the path. The formula for this case is:

$$\text{Distributed Float} = \text{Activity Duration} / \text{Path Duration} \times \text{Total Float}$$

So the distributed float of activity #1 is: $3/24 \times 6 = 0.75$ days; for activity #2, the distributed float is: $21/24 \times 24 = 5.25$ days. Thus, the new Late Finish for activity

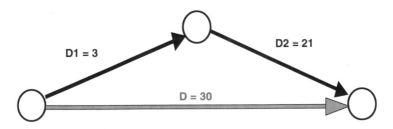

Fig. 8.22 Backward pass allows for calculating late start and finish dates of activities

#1 would be 3.75 days and the new Early Start for activity #2 would be 24.75 days (30 days, which is the project duration, minus the distributed float).

Another way to distribute the total float is by using a float-sensitive distribution. That is, considering the length of the activities as well as the inherent risk in the activity itself. The formula in this case would be:

$$\text{Distributed Float} = \text{average (Activity Duration /Path Duration;}$$
$$f\,(\text{risk})) \times \text{ Total Float}$$

In the example above, if activity #1 is the design phase of a project, with an 80% risk of delay, and activity #2 is the construction phase of the same project, with a 20% risk of delaying the project then:

$$DF(A1) = \text{average } (3/24;\ 0.8)\ \times\ 6 = 2.775$$

$$DF(A2) = \text{average } (21/24;\ 0.2)\ \times\ 6 = 3.225$$

So, the new Late Finish for activity #1 is 5.775 days, whereas the new Early Finish for activity #2 is 26.775 days (30 days minus the distribution float).

8.5 Precedence Diagramming Method

As previously discussed, PDM is an AON network method and goes beyond the CPM by including other inter-activities relationships such as Start-to-Start (SS), Start-to-Finish (SF) and Finish-to-Finish (FF) apart from the conventional Finish-to-Start (FS).

It also includes the possibility of adding "lags" or "leads" (negative "lags") between activities. If we consider that there is a relationship XY (SS, SF, FS or FF) with lag "t" between activities A and B, then event Y of activity B can occur no earlier than t units after event X occurs for activity A.

Figure 8.23 illustrates different situations of leads and lags.

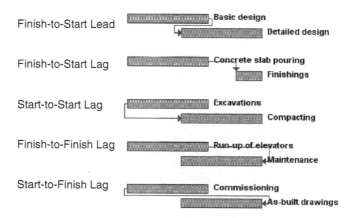

Fig. 8.23 Relationships, lags and leads in a CPM schedule

Fig. 8.24 Example of a reverse critical activity

Nodes now are no longer simply vertices in the graph. Here, an arrow on the left side of the node indicates a Start Relationship, whereas an arrow on the right side of the node indicates a Finish relationship.

In the PDM, the user can also add some constraints as in the CPM by assigning a fix date to a particular activity (it works as a milestone). One just needs to remember that milestones are given priority over relationships or other kind of links, so pay special attention to give "reachable" milestones. Otherwise, the links one may propose will be broken. Also, the user can set dates under the form of "must start/finish" constraints or as-late-as-possible calculations (e.g. must start on, no early than, etc.).

Some caveat of PDM need to be pointed out. It is important that the user clearly understands all the different relationships between activities, especially concerning the "lead" and "lag" concepts, which usually lack a specific standard and change from software to software. It is also important to stress that for a same activity there may be two differing floats: the Start Float (Late Start – Early Start) and the Finish Float (Late Finish – Early Finish).

As far as the CP under a PDM notation is concerned, choices on the relationships between activities clearly impact the critical path and tracing the critical path may be difficult for various reasons. For example, non-critical activities may have a critical start or finish date. Also, the critical path of the network may go backward through an activity, with the result that increasing the activity time may actually decrease the project completion time. Such an activity is called "reverse critical" and this happens when the critical path enters the completion of an activity through a finish constraint, continues backward through the activity, and leaves through a start constraint, as in the example drawn in Fig. 8.24 (the longer Activity 2 is, the smaller the critical path duration – and the quicker the project can be completed):

Furthermore, as far as different software packages display the critical path differently, it is of great importance for the scheduler to use the software package as a tool and not to completely rely on its outcomes (e.g.: Microsoft Project displays as-late-as-possible constrained activities as critical if the project is scheduled from the start date).

8.6 Resource-Based Scheduling

This section completes the CPM technique, according to which an optimal duration can be determined as a result of the optimization of the time-cost tradeoff. This is still an open problem and involves the application of heuristic algorithms to find the minimum total cost consistent with the project optimal duration.

This section also addresses situations that involve resource optimization. Resource-constrained scheduling applies whenever there are limited resources available and the competition for these resources among the project activities is keen. In short, the time-cost optimized schedule can provide a bad utilization of resources with high peaks and under loaded periods. Resource leveling aims to minimize the period-by-period variations in resource loading by shifting tasks within the allowed slacks.

Another problem is about resolving periods with over allocated resources: heuristic models require priority rules to establish which activity takes precedence in resource usage and which one can be postponed or get a longer duration.

8.6.1 Time-Cost Schedule Optimization with CPM

Let us recall the critical path method: once activities are defined from WBS work packages and durations for each activity as well as cost and resources are estimated, then it is possible to plot the network and perform the CPM scheduling to estimate time, cost, and resource usage over the whole project.

If the total duration is compliant with the contract baseline, the schedule is terminated. If it is not acceptable, it is needed to impose other dependencies or added resources in order to reduce the project total duration (*"project crashing"*). Indeed, so far, scheduling has been referred to as time allocation; but, since time is a function of resource usage and the inherent related cost, possible tradeoffs exist between time and cost, and, more generally, between time and resources.

There are several ways to crash a project: supplying a higher number of human resources, using overtime or multiple shifts, and changing the technology.

Adding additional resources may not be possible or effective for several reasons. First, the available supply of a limited resource might be exhausted. Second, the wage for addition resources may be higher, or the resources might come in packages, such as a crew of 3 electricians. Thirdly, the productivity of additional resources might not be as high as the original resources. Training may be required, or limitations such as space or the nature of the task at hand might cause a slowdown of work.

Increasing the number of shifts avoids the problem of reduced productivity die to crowding, but has problems of its own, such as the increased cost of labor at night, and the natural fact that people are less productive over night.

Overtime is an option, but worker productivity drops dramatically after 40 h a week. Productivity rebounds slightly for a few weeks, but then drops off again. Overtime wages are also more costly than standard wages.

A change in technology can also reduce time and costs, but also has some drawbacks. More efficient equipment is most likely more expensive. Changing technology in a project might also create the need for some redesign or rework. In the end, the time saved by a technology change might not be linear to the additional costs incurred. One thing that needs to be considered when scheduling is the type of task at hand. If the task has a fixed duration, such as the curing time of concrete, it cannot be crashed. In order to save time in this scenario, the technology might have to be changes to quick setting concrete.

As a result, project crashing inevitably increases the cost of the project: we call "crashed", or accelerated, cost the cost associated with a crashed, or accelerated, duration of the network.

With this notion in hand, it is possible to optimize the network using the CPM.

The first task is to schedule the project using a "normal" time frame and associated "normal" cost. The second step is to crash the project. This is done for two reasons: to reduce the normal finish date to less than the contract deadline if needed, and to establish the length of the project at minimum costs.

Crashing a project consists of reducing the time that it takes to complete the project. Usually this raises the cost of the project.

In most cases, there are a few portions of the project that can be crashed, resulting in a high reduction in project time, but relatively small increases in cost. As more and more tasks get crashed, the relative gain in time to the increase in costs gets smaller. At some point it is no longer valuable to trade time for costs. The chart of Fig. 8.25 gives an example of this.

When crashing a project, it is important to look at the critical path. There is no reason to crash tasks not on the critical path, because no time on the project will be saved, resulting in more cost with no time benefit. It is also important to watch how the critical path changes during crashing. After crashing a few tasks, the critical path might change, and then tasks not originally on the critical path will need to be crashed to reduce the project time. This makes a big difference in construction project management, as many managers would crash all the tasks in a project to save time, when it is unnecessary to crash many of the tasks, as in the R point illustrated in Fig. 8.26.

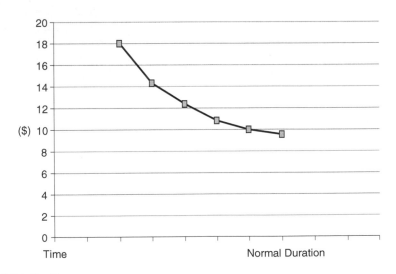

Fig. 8.25 Crashing curve

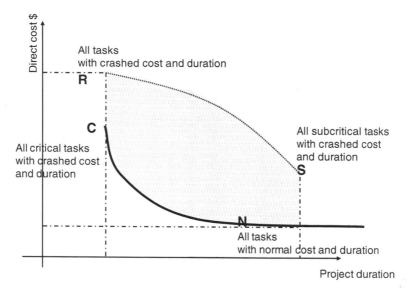

Fig. 8.26 Time-cost configuration space resulting from all possible crashings of the project duration

More generally, Fig. 8.26 shows the time-cost configuration space resulting from all possible crashing simulations. It illustrates the importance of speeding up only the tasks that are on the critical path, since there may be ineffective situations when a shorter duration of the project may be obtained with lower cost (e.g. point R versus point C of the graph).

In other words, the proper "crashing curve" is the one that is not dominated by more efficient curves: the crashing curve is a Pareto-optimal solution because it minimizes the direct cost associated with a given duration of the project. Also, it is worth to note that the crashing curve slope increases as the duration is crashed up to its shortest date; in fact, crashing is limited by technology and resources to a minimum duration. On the opposite side, a longer duration than the one calculated with a minimum/normal usage of resources does not lead to lower costs.

At this point, it is convenient to take a look at the curve of the total cost, which sums the direct cost with project and corporate overhead costs. Because overhead increases as the project continues off, there is a minimum point of the total cost curve, as shown in Fig. 8.27. The minimum point determines the *"optimal duration"* referred to as the length of the project consistent with the minimum total cost for the firm.

A better insight of the graph above suggest that there may be reasons to pay for penalties up to the optimal duration of the project: from the date when contract penalties are due to the optimal finish date it is less expensive to sustain overhead and pay the liquidated damages than to afford the cost of crashing. In general, later than the optimal duration the daily crashing cost is less than daily overhead.

The optimal duration allows define the proper amount of crashing. But, whether or not accelerating the entire project, crashing is suitable for different algorithms. In

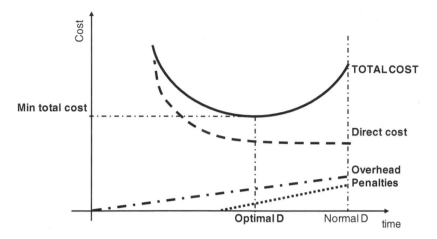

Fig. 8.27 Optimal duration is the one who minimizes the total cost of the project

particular, during the planning phase it may be simply opportune to crash the initial activities of the critical path, while during the project execution there is no other possibility than crash the remaining activities, from time now forward. In any case, the definition of a proper crashing algorithm is required.

If activity time-cost curves are linear, then finding the optimal duration of the project is a linear programming problem. Unfortunately, in most cases there is no straight-linear relationship between time and cost. This ends up to a non-linear programming problem for which the definition of an algorithm based on heuristics is required.

As a general rule, basic recommendations apply to schedule crashing heuristics:

- focus on critical path, which means that only critical activities should be crashed (note that as the crash time amount increases, the number of critical activities increases as well);
- select the less expensive way to do it, that is crash first activities that result in a smaller increase in costs
- trade time for money on non-critical activities: the activity time should be lengthened to reduce costs if possible. Non-critical paths can be extended within the available float, reducing the costs of the task. As long as the task is not extended beyond the available float, the project duration is not lengthened, and the indirect costs will not go up.

One of the most used crashing heuristic algorithms is the one by Kelly and Walker (1959). It states the following steps:

1. solve CPM with normal durations;
2. for critical activities find marginal cost of crashing (i.e., additional cost of shortening duration 1 time unit);

3. reduce by one time step the critical activity with the lowest marginal cost of crashing;
4. record resulting project duration and cost;
5. repeat step 3 until another path becomes critical.

8.6.2 Resource Leveling

Mainly, limitations to the schedule regard the tradeoff cost-time-resources. If the project is budget limited, then it must have duration and resource usage based on the required preset cost. In this case, resources must be leveled to reduce indirect costs and then total costs. In a time limited project, it must finish within a scheduled date thus requiring resource usage at best with minimum possible cost.

Finally, if a project is resource limited, it must not exceed a specific level of resource usage or overcome resource constraints (such as crew sequencing) so that the project duration is the shortest possible time associated with the limitation. If the normal schedule was based on some resource limitations, then crashing the schedule may not be possible, or might greatly increase costs due to working around the resource limitations. In short, the time-cost optimized schedule can suggest an impossible schedule or provide a bad utilization of resources with high peaks and under loaded periods.

To solve the problem, *"resource leveling"* may help to reduce the period-by-period oscillations in resource loading by shifting tasks within the allowed float. During the course of a project, a more steady usage of resources leads to lower resource costs. This is due to reducing the costs of hiring, training, and firing human labor, material storage, and equipment rental and storage. Resource leveling is done by moving tasks around in the schedule and reorganizing the floats.

Figure 8.28 shows an example of this. The original resource-load profile (dashed line profile) can be leveled if the non-critical activity #2 is anticipated (black-marked profile). This provides a double advantage: it avoids a later short resource peak and keeps the maximum amount of resources within 25 units over the project.

In some cases the situation arises where the resources cannot be leveled within the available float. At that point, a decision needs to be based on what is more cost-effective, whether acquiring more resources or lengthening the schedule. When resources are limited, the schedule has to be lengthened or reworked to accommodate the situation. One thing to keep in mind is that performing tasks when possible is not always the best approach. When resources are limited, the entire project schedule needs to be evaluated to find the optimal way to work around the constraints.

The following is an example of "manual workaround" (i.e. adding precedence relationships to the original network links) that is needed to respect resource constraints. Suppose that there is only one crane available to perform two overlapping activities requiring a crane. The solution is either to add a precedence link between

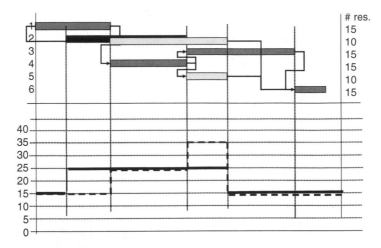

Fig. 8.28 Resource leveling within available floats

the two activities to have them performed in sequence (with twice the original duration of the path) or to buy/rent an extra crane with added cost.

8.6.3 Heuristic Scheduling Approaches

A wider aspect of resource scheduling is concerned with resource leveling under limited-resource allocation. This combined problem can be solved either trough optimization or by applying heuristics algorithms.

Finding the "optimal" configuration for leveling a resource-constrained schedule is a computationally-expensive combinatorial problem. In principle, it would need to compare all possible orderings of conflicting activities. Applications on specific projects exist based on approaches such as linear programming, explicit enumeration and "Branch and Bound" methods.

Heuristics algorithms, though inconsistent with finding the optimal allocation, yet provide useful configuration of leveled resource-constrained schedules. Heuristics use some "rules of thumb" to get answer in an acceptable time. They typically reach a local minimum and do what is locally-best, but not necessarily globally-best. This means that heuristics may not optimize the project as a whole.

There are two types of resource-scheduling heuristics. The "serial methods" schedule activity-by-activity: the algorithm considers prioritized activities in order and schedules them as early as possible. Activities are assigned priority based on a number of attributes: length, resources required, slack, or the number or type of successors. Each activity needs to wait until its predecessors have been completed, and the required number of resources is available.

The "parallel methods" schedule activities by time step. At each time step, some activities are delayed as needed based on algorithm criteria such as Shortest Task

First or Longest Task First and rules like: "Wait until predecessors complete or adequate resources are available".

Parallel methods are more commonly used in current software packages (i.e. Microsoft Project, Primavera) than serial methods.

References and Additional Resources About Planning and Scheduling

Diamant L, Tumblin CR (1990) Construction cost estimates. Wiley, New York, NY

Fisk ER (2003) Construction project administration, 7th edn. Pearson Prentice Hall, Upper Saddle River, NJ

Hinze JW (2004) Construction planning and scheduling, 2nd edn. Pearson Prentice Hall, Upper Saddle River, NJ

Kelly JE, Walker MR (1959) Critical path planning and scheduling. Mauchly Associates, Ambler, PA

Kerzner H (2001) Project management: a systems approach to planning, scheduling, and controlling. Wiley, New York, NY

Meredith J, Mantel S (2006) Project management: a managerial approach, 6th edn. Wiley, New York, NY

Moder J, Phillips CR, Davis EW (1995) Project management with CPM, PERT, and precedence diagramming, 3rd edn. Van Nostrand Reinhold, New York, NY

Associated General Contractors of America (2005) Construction estimating & bidding: theory/principles/process. AGC of America, Arlington, VA

Patrick C (2004) Construction project planning and scheduling. Pearson Prentice Hall, Upper Saddle River, NJ

Chapter 9
Project Monitoring and Control

9.1 The Monitoring Process

Given the little likelihood that the project will remain firmly on schedule and at cost, a project manager has to reduce cost overruns and time delays to a minimum, through effective Project Monitoring. This is a management method to measure the project actual progress and cost and time current performance.

An effective measurement system is a basic requirement for controlling quality, cost and time. Based on the results of Project Monitoring, the project team can then activate a Project Control process to ameliorate any issues and return the project to its scheduled course (Ritz 1994). Project Control is a recurring process involving comparison of actual performance to scheduled performance, estimates to completion and corrective actions based on such estimates, which often require either performance adjustments or schedule revision. Monitoring and Control (often shortly called Project Control) are two parts of a feedback system aimed at detecting and correcting deviation from desired (Fig. 9.1). Detection is made through monitoring, while correction is the objective of control actions.

Monitoring can be defined as the set of procedures and management practices used to collect information about the performance achieved or forecasted in a project, based on a set of performance metrics. Monitoring includes performance analysis of the project, which is the process of determining performance variances based on monitored and forecasted performance. Control adjusts the project to meet its initial goals by analyzing the causes of performance problems, designing changes to address problems that are determined to need attentions, and implementing those changes through control actions (MIT Open Courseware).

This chapter presents the most used tools and techniques for project monitoring and how this is paired with Project Control to actually affect project performance.

9.2 Measurement of Actual Progress

Project monitoring is a process that must be carried out all along the project execution. It is aimed at identifying deviations from an existing plan and giving real-time information to allow for making appropriate project control policies and decisions.

A. De Marco, *Project Management for Facility Constructions*,
DOI 10.1007/978-3-642-17092-8_9, © Springer-Verlag Berlin Heidelberg 2011

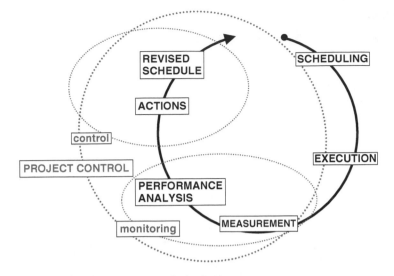

Fig. 9.1 Monitoring and control as part of a feedback system

Important components of an effective monitoring system are a detailed WBS and schedule for accurate measurement of job progress, the establishment at the planning phase of a useful representative performance metrics, a management scheme organized for honestly and accurately identifying and reporting of performance, the involvement of responsible and knowledgeable people in the reporting scheme, as well as project reviews, meetings, inspections, and audits (Ritz 1994; Meredith and Mantel 2006).

The first aim of monitoring is to track real-time project progress. This must be done based on the predetermined WBS/CBS system: specific items in the detailed WBS are designated as job cost accounts (i.e. the CBS is the same as the WBS) for recording of expenses and comparison to the baseline cost of each element. This method allows for accounting based on activities and permit tracking expenditure by activity and work item (Yang et al. 2007).

Sometimes, a different cost structure from the WBS is required to be monitored, as in the case of an existing corporate cost control system. Recording a cost granularity different from the activity breakdown is critical (Jung and Woo 2004). This requires a many-to-many mapping between WBS activities and CBS cost categories thus complicating the progress measurement process for foremen and superintendents (see 7).

Regardless the accounting system used, project monitoring is based on the following main steps:

1. measurement of actual cost and schedule progress,
2. calculation of the discrepancy between actual status versus scheduled progress (trend),
3. estimation of cost and time at completion of the project (based on trend).

Leveraging the construction budget and current cost statements can give a general perspective on project progress. Simply comparing actual expenditure against the budget, however, is of limited and tricky use, as shown in the following example.

Example – Measurement of Progress Based on Actual Cost Vs. Budget

If you are managing a project with the following characteristics:

Duration = 18 months
Budget = $1,100,000

and at 10 months into the project:

Actual consumed cost = $700,000

the project progress estimated as:

Actual cost/budget = 700,000/1,100,000 = 63.63%

is logically flawed and highly inaccurate. In fact, the project could easily be over projected costs and behind schedule but this technique is too simple to inform the project manager of actual physical progress. Here, approximately 63% is the portion of budgeted cost actually spent after 10 months: this does not give information on the amount of work done related to that actual expenditure.

Thus, to determine the project progress it is necessary to activate the process of determining the actual progress of the work physical output of individual item/activity of the WBS.

Several are the metrics to evaluate the physical progress of individual activities (U.S. DoE 1980; Eldin 1989, Fleming 1992) depending on the kind of task.

In the case of tasks that involve production of easily measured deliverables, the "units completed" are practicable and viable metrics for assessing the actual situation based on the measurement of physically completed units (e.g.: m^3 concrete poured, etc.). In such a process of determining the job done, foremen and superintendents are often required to record performed quantities in Quantities Book and to enter actual worked hours in labor Timesheets. Additional managerial attention may be paid on important activities and special items.

Accordingly, individual activity progress is defined as the percent ratio of actual performed quantity over total scheduled quantity:

Progress = performed quantity/total scheduled quantity [lb, kg, ft, m, etc.]   (9.1)

Following is an example of how individual activity progress can be counted based on actual quantity recording. The example is about the construction of a pre-cast concrete warehouse.

In Table 9.1, the percent progress can be more than 100% if performed units are more than the planned ones. This is appropriate if the project is executed under the conditions of a contract that does not allow for change orders. If the

Table 9.1 Measurement of activity progress

WBS	Unit	Actual quantity	Scheduled quantity	Activity progress (%)
Structures				
Footings				
Procurement	Unit	79	76	103.95
Shipping	Unit	79	76	103.95
Construction	Unit	79	76	103.95
Columns				
Procurement	Linear meter (lm)	108	220	49.09
Shipping	lm	108	220	49.09
Erection	lm	108	220	49.09

additional units are due to a reimbursable change, it may be worth managing the extra work as a separate project with autonomous WBS, budget and progress recording.

Sometimes, the task of measuring percent progress based on quantities is more difficult: this is the case of activities producing a nonmaterial output, such as design and engineering, where the production input is different from the output. For example, measuring the progress of hours spent in design may not reflect the actual status of drawing advancement. The challenge in progress estimation is to measure the output progress, so that agreed target-based progress measurement metrics are required.

The "on/off" technique is a useful approach when accounts cannot be physically measured: progress of an item is 100% complete when the job is formally accepted as finished. Alternatively, the "0-50-100%" metrics allows for recording future, underway and completed tasks.

Since with the on/off techniques the progress record of underway tasks is nil, the method may not result in an underestimation of performance only if measurement is carried out at the very detailed level of small elementary tasks, which may be not always the case. To overcome the problem, different conventional metrics may be established as a set of "incremental milestones", such as the ones required to preparing, submitting and approving a document, as in the example presented in Table 9.2.

In this case, conventional percentages are associated to each step of the process based on the number of work-hours, or other quantities, estimated to be required to that point in relation to the total. The incremental milestone approach well applies to longer activities and to the measurement of WBS items at a higher level than the detailed task one.

Other difficult-to-measure activities are those that take place in a variety of places which make physical observation and measurement hard or that require the execution of preliminary activities before the work in place. In these cases, schedulers and project managers are often forced to rely on subjective judgments risking in under

Table 9.2 Example of contract-agreed metrics to measure progress of design and engineering activities

Type A1 Doc. for owner approval	Progress (%)	Type A2 Doc. internal only	Progress (%)	Type A3 Mat. Req. Bid Eval.	Progress (%)
Start, studies collect	20	Start, studies collect	20	Start, studies collect	20
First issue	60	First issue	70	Bid	75
Returned comments	70	Final issue RFC	100	Order issued	100
Second issue	80				
Returned approved	90				
Final issue RFC	100				

Table 9.3 Use the original budget to assign cost-based weights to each project activity

WBS	Unit	Total quantity	Unit cost [$]	Budget [$]	Weigh (%)
Structures				1,100,000	100.00
Footings				148,200	13.47
Procurement	Unit	76	1,400.00	106,400	71.79
Shipping	Unit	76	200.00	15,200	10.26
Construction	Unit	76	350.00	26,600	17.95
Columns				951,801	86.53
Procurement	Linear meter (lm)	220	2,800.00	616,000	64.72
Shipping	lm	220	326.37	71,801	7.54
Erection	lm	220	1,200.00	264,000	27.74

or overestimation of work done. In such circumstances, it is recommended to refer job progress to a reference parameter or to breakdown the activity into more detail and measure those details.

Progress of indirect cost may be assessed in terms of apportioned effort in relation to the progress of activities they are linked to.

Of course, it is possible to use other methods to measure actual progress of individual work packages and tasks. In any case, a flexible use of all methods permits that each item is recorded in the more reliable way.

Finally, to estimate the overall project progress, one method is to use the Original Budget to give cost-based weights to each project activity (Table 9.3) and calculate the overall percent progress as a weighted sum of individual activity percent progress (Table 9.4).

Briefly, in the example above, the physical progress, based on performed quantities, of the overall project progress is approximately 56.5%, while the cost consumption attains 63.6%, as calculated in the Example box above and confirmed by the cost report in the following Table 9.5.

Table 9.4 Calculation of the overall project progress as a weighted sum of all task progress measurements (from Table 9.1)

WBS	Unit	Actual quantity	Scheduled quantity	Activity progress (%)	Weigh (%)	Project progress (%)
Structures					100.00	56.48
Footings					13.47	14.00
Procurement	Unit	79	76	103.95	71.79	74.63
Shipping	Unit	79	76	103.95	10.26	10.66
Construction	Unit	79	76	103.95	17.95	18.66
Columns					86.53	42.48
Procurement	Linear meter (lm)	108	220	49.09	64.72	31.77
Shipping	lm	108	220	49.09	7.54	3.70
Erection	lm	108	220	49.09	27.74	13.62

Table 9.5 Actual cost report at 10 months into the project

WBS	Unit	Actual quantity	Unit cost [$]	Actual cost at month 10 [$]
Structures				700,001
Footings				162,209
Procurement	Unit	79	1,500.00	118,500
Shipping	Unit	79	203.28	16,059
Construction	Unit	79	350.00	27,650
Columns				537,791
Procurement	Linear meter (lm)	108	3,200.00	345,600
Shipping	lm	108	400.00	43,200
Erection	lm	108	1,379.55	148,991

9.3 Performance Measurement: Earned Value Analysis

Now, the challenge here is to measure and forecast the project cost and time performance using monetary information. As discussed, the problem of traditional comparison between actual cost versus scheduled cost does not take into account the progress status of the project.

Earned Value Analysis (Project Management Institute 2008b) is an extremely effective way to overcome the problem. Earned Value Analysis (EVA) integrates cost, schedule, and work performed by ascribing monetary values to each. EVA is a method for measuring project performance.

Earned Value Analysis is based on three key values:

- BCWS (Budgeted Cost of Work Scheduled) is the planned cost of work scheduled to be accomplished in a given period of time;
- ACWP (Actual Cost of Work Performed) is the cost actually incurred in accomplishing the work performed within the control time;
- BCWP (Budgeted Cost of Work Performed) is called Earned Value. It is the budget value of the work actually performed within the control time.

Table 9.6 Earned Value at 10 months into the project execution

WBS	Unit	Actual quantity	Unit budgeted cost [$]	Earned value at month 10 [$]
Structures				621,298
Footings				154,050
Procurement	Unit	79	1,400.00	110,600
Shipping	Unit	79	200.00	15,800
Construction	Unit	79	350.00	27,650
Columns				467,248
Procurement	Linear meter (lm)	108	2,800.00	302,400
Shipping	lm	108	326.37	35,248
Erection	lm	108	1,200.00	129,600

Earned Value is the budgeted value of the work completed to date. EVA simply compares this amount to the actual cost of work completed to understand cost discrepancies and to the budgeted cost of work scheduled to assess any schedule variances.

Table 9.6 is the analytic calculation of the Earned Value of each activity in the example considered

The same result may be obtained through synthetic determination of the overall project earned value, as below:

$$EV = BC \times WP = \$1,100,000 \times 56.48\% = \$621,280 \qquad (9.2)$$

As mentioned, traditional comparisons of actual cost versus budget fail in considering the amount of work done. The Resource Flow Variance (RV) or the Resource Flow Index (RI) compare how much expecting to spend during a timeframe with what actually spent, regardless of how much work got done. They are defined as:

$$RV = BCWS - ACWP \qquad (9.3a)$$

$$RI = BCWS/ACWP \qquad (9.3b)$$

They do not indicate a bad or good situation; for example, a project may go faster but more cheaply than expected or go slower but more expensively than expected.

Earned Value Analysis incorporates a number of derived metrics to get a better feel with regard to project performance.

One metric is the Cost Variance (CV), the difference between the Budgeted Cost of Work Performed and the Actual Cost of Work Performed, or the corresponding Cost Performance Index (CI), as the ratio between the two values:

$$CV = BCWP - ACWP \text{ (earned value } - \text{ actual value)} \qquad (9.4a)$$

$$CI = BCWP/ACWP \qquad (9.4b)$$

This serves as a comparison of the budgeted cost of work performed with the actual cost incurred. A positive value means the project is underrun, with a gain of value. A negative variance means the project is over budget (loss of value). A zero value means the project is on budget. Likewise, a CI of less than one means project is overrun, while a CI more than one indicates a budget underrun.

Another is the Schedule Variance (SV), which is the difference between the Budgeted Cost of Work Performed and the Budgeted Cost of Work Scheduled or the corresponding quotient Schedule Performance Index (SI):

$$SV = BCWP - BCWS \text{ (earned value } - \text{ budget value)} \qquad (9.5a)$$

$$SI = BCWP/BCWS \qquad (9.5b)$$

This gives a comparison of the amount of work performed during a given period of time to what was scheduled to be performed. A negative schedule variance means the project is behind schedule (loss of time), a zero value suggests the project is on schedule, and a positive SV means the project is ahead of schedule (gain of time). Similarly, SI less than one is behind schedule, equal to 0 is on schedule, and more than one is ahead of schedule.

Example – Synthetic EVA

Consider the case-example discussed throughout this chapter and suppose that the Budget Value at 10 month into the project was scheduled to be $660,000 corresponding to 60% of progress.
Recalling of the figures for the overall project provides the following:

- Budget at Completion (BAC) = $1,100,000
- Budget Value (BV) = BCWS = $660,000
- Actual Value (AV) = ACWP = $700,000
- Earned Value (EV) = BCWP = $621,280

It is possible to compute the variances of the overall project:

- Resource Flow Variance = BV – AV = 660,000–700,000 = –40,000$
- Cost Variance = EV – AV = 621,280–700,000 = –78,720$

Note that $40,000 are the accounted extra costs consumed to reach 56.48% progress, where the minus sign means a loss of money. Although, $78,720 are the costs that will be consumed to attain the 60% progress that was scheduled to be performed within 10 months. Here also the minus sign indicates loss of value.
Also, the schedule variance is:

– SV = 621,280–660,000 = –38,720$

that is exactly the difference between the cost and the resource flow variance. This tells that the project is consuming $40,000 as more actual cost than estimated, as well as $38,720 to perform the schedule delay.

Synthetic calculation of variances for the overall project, using weighted sums of progress and overall values, is of great convenience to the project manager for communication and rapid reporting.

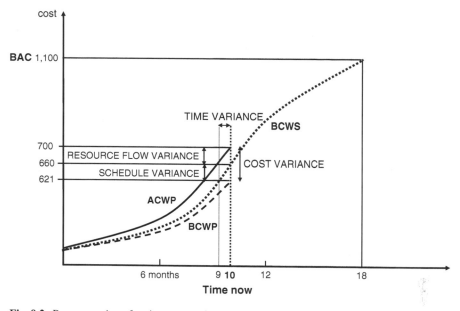

Fig. 9.2 Representation of variances on a S-curve chart

Also, resource, cost and schedule variances can be represented in a time/cost chart using S-curves of BCWS (dotted line in Fig. 9.2), ACWP (dashed line in Fig. 9.2) and BCWP (solid line in Fig. 9.2). This enables a quick graphical understanding of the project status for suggestion of global corrective actions to the project strategy.

Planned (Scheduled), Actual and Earned Value S-curves can have six possible arrangements, as in the chart presented with Fig. 9.3.

One must look at the position of the EV S-curve as the reference curve line: whether the AV or the PV above the EV curve indicates the project is over budget or is behind schedule. The best case-scenario is the one where both the AV and the BV curves are below the EV: in this case both the cost and the duration are performing better than scheduled. The more the distance of the AV and BV curve lines from the PV one, the larger the loss or gain of value and schedule.

Close reasoning can be conducted for Cost and Schedule performance indexes: aggregate indexes can be obtained for the overall project by weighted sum of activity performance indexes. Figure 9.4 is a chart showing that the best case-scenario is when both the aggregate CI and aggregate SI for a project are more than 1.

Whereas a synthetic report of the overall project performance prevent from information overload to direct the project's decision-makers to the major issues, analytic EVA provides greater details to allow a better understanding of the job status through investigation of individual activity performance. A detailed report based on activity breakdown informs the project manager about the parts of a job that are behind schedule and over budget and helps in finding out why the job is in its current status.

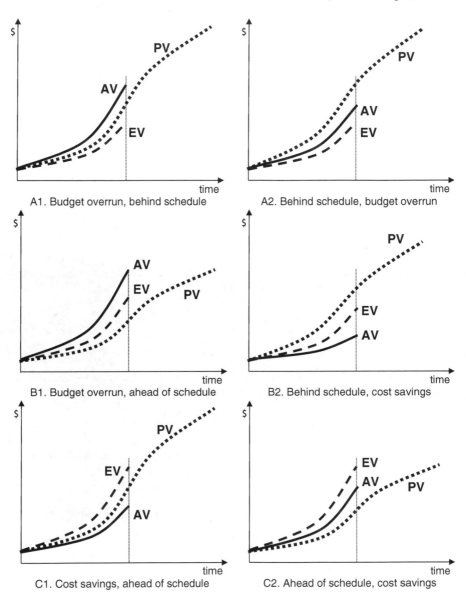

Fig. 9.3 Possible arrangements of S-curves indicating planned value (PV), actual cost (AV) and earned value (EV). For example, A1 indicates both cost overruns and schedule delay, with more serious problems on cost than on schedule. A2 is a similar situation, where schedule delay is more significant than extra cost

The Cost Variance Report for the example-project is given in Table 9.7. It is worth noting that the activity "Columns procurement" is the only one recording a positive value of the RV. This is pretty tricky: indeed, that task is the one with the biggest problems (a negative CV of $43,200) with both an increased unit cost and a schedule

Fig. 9.4 Areas of cost and time performance using cost and schedule performance indexes

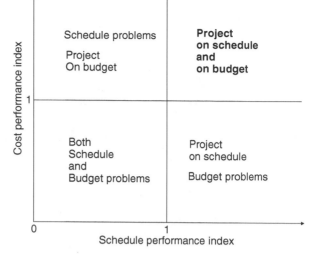

Table 9.7   Example of earned value report

WBS	WS (%)	BV (BCWS)	WP (%)	AV (ACWP)	EV (ECWP)	BV-AV	EV-AV
Structures	60.0	660,000	56.48	700,001	621,298	38,702	(78,703)
Footings	13.0	148,200	14.00	162,209	154,050	(5,850)	(8,159)
Procurement	100.0	106,400	74.63	118,500	110,600	(4,200)	(7,900)
Shipping	100.0	15,200	10.66	16,059	15,800	(600)	(259)
Construction	100.0	26,600	18.66	27,650	27,650	(1,050)	—
Columns	47.0	511,800	42.48	537,791	467,248	44,552	(70,543)
Procurement	65.0	100,100	31.77	345,600	302,400	98,000	(43,200)
Shipping	45.0	32,311	3.70	43,200	35,248	(2,937)	(7,952)
Erection	30.0	79,089	13.62	148,991	129,600	(50,511)	(19,391)

delay. The reason may be found in the procurement process of columns as a whole: it seems that the pre-cast columns manufacturer stopped to provide materials up to a 49% of units.

A timely performance reporting would have been highlighted the supply problem ahead of time to have the project team tightly work with the supplier and avoid the inconveniency.

9.4   Forecasting Performance

The CV and the SV, as well as the CI and the SI are factors of past behavior to use as trends for predicting future targets, if no corrective actions are undertaken (CII 2004). Thus, the cost estimate at completion and the time estimate at completion can be calculated by extrapolating the actual performance to the end of the project (Project Management Institute 2008a).

Forecasted cost is based on the following principle:

Forecasted cost = Cost spent + (Work remaining * Expected unit cost), where:
Expected unit cost = Costs spent/Work performed

Using data available from an EVA report, it is possible to calculate Estimate at Completion (EAC) in a couple of ways. The first original approach states that future remaining cost will be in line with the budget (i.e. the total Budget at Completion minus the budgeted cost of work performed). This means:

$$EAC = ACWP + (BAC - BCWP) = BAC - (BCWP - ACWP) = BAC - CV$$
$$(9.6)$$

This approach is rather optimistic, assuming that cost overruns are old problems and will not incur in the future. A better way for calculating EAC is a revised estimate approach:

$$EAC = ACWP + (BAC - BCWP)/CI = BAC/CI \qquad (9.7)$$

This principle assumes that the project future will, at least, reflect the past performance, if no corrective actions are undertaken.

Example – Cost Estimates to Completion

Consider the case-example; we can calculate the following EAC.
Original estimate approach:
    EAC = BAC – CV = 1,100,000–(621,000–700,000) = $1,179,000
resulting in a Variance at Completion (VAC) = –79,000
Revised estimate approach:
    EAC = BAC/CI = BAC × (ACWP/BCWP) = 1,100,000 × (700,000/621,000) = $1,240,000
    VAC = 1,100,000–1,240,000 = – 140,000

Similarly, forecasted completion dates are based on the following principle.

Actual Completion date = Current date + (Work remaining/Expected work rate)

Using data available from an EVA report, it is possible to calculate the Actual Completion date (AC) according to either an original or a revised approach. The original estimate approach assumes that time overruns are past history and will not incur in the future, so that:

$$AC = \text{Current date} + (\text{Work remaining/Scheduled work rate})$$
$$(BC - T)$$

$$AC = T + (BAC - BCWP) \times \text{------} \atop (BAC - BCWS) \qquad (9.8)$$

With the revised estimate approach this is calculated as:

$$AC = T + (BAC - BCWP) \times \frac{(BC - T)}{\text{--------}} = BC/SI \qquad (9.9)$$
$$(BAC - BCWS) \times SI$$

In the formulae above T is the current date (time now) and BC is the scheduled completion date.

Example – Time Estimates at Completion

Consider the case-example; at 10 months into the project, with a scheduled total duration of the project of 18 months, it is possible to calculate the following actual completions.

Original estimate approach:
$$AC = 10 + (1,100,000 - 621,000) \times (18 - 10)/(1,100,000 - 660,000)$$
$$= 10 + 479,000 \times 8/440,000 = 10 + 8.71 = 18 \text{ months} + 22 \text{ days}$$

Time overrun = 22 days

Revise estimate approach
$$AC = 10 + (1,100,000 - 621,000) \times (18 - 10)/[(1,100,000 - 660,000)$$
$$\times (621,000/660,000)]$$
$$= 10 + 479,000 \times 8/414,000 = 10 + 9.26 = 19 \text{ months} + 8 \text{ days}$$

Time overrun = 38 days

Some corrections in Eq. (9.7) may help in better predicting the cost at completion. Christensen (1999) remarks that a different performance factor may be used to account for the integrated influence of the schedule variance to the cost performance: a bad SI may be an indicator of future cost overruns. Thus, the performance index for estimating the cost at completion may be either the CI or the SI, or some combination of the two (e.g.: CI × SI; 0.8 CI + 0.2 SI). The diverse adjustments lead to a range of EACs, where the CI factor calculation is a reasonable floor to the final cost, and the one obtained by using the product of CI and SI is a sufficiently large indication of the maximum final cost.

Also, Lipke (2005) has experimentally determined that in engineering projects a correction factor to the CI allows for calculating the EAC upper bound. The maximum correction factor equals 0.1 and Eq. (9.7) may be rewritten as:

$$BAC/CI < EAC < BAC/(CI - 0.1) \qquad (9.10)$$

Alternative way to forecast cost performance is to use the cost per progress percent point indexes. This method assumes that advancement of project is linear and allows defining:

actual cost per progress point = Actual value/work performed = ACWP/WP

$$(9.11)$$

Then, to calculate cost and duration to completion, simply bring the unit index to 100 points.

Example – EAC Using Percent Indexes

Cost at completion are calculated as:

EAC = ACWP/WP * 100 points = 700,000/56,48 * 100 = 1,240 * 100
    = 1,240,000

If we recall that the scheduled cost per progress point is 1,100, the cost overrun per progress point is worth $140, which means that the project cumulates losses of value at a rate of $140.

With sensible variation, the index-based formulas above, which assume that the project progresses strait-linearly, allow in any case for an accurate evaluation of EAC and alternative nonlinear models, such as logistic curve fitting methods, are not as practicable compared to the little additional precision they provide (Christensen et al. 1995).

Yet, index-based estimates at completion have some restrictions. In fact, the linear model assumes that the latest measured performance will remain the same until the project is complete without taking into account any late performance change that typically apply to construction projects. Also, the elements used for progress measurement must be broken down with a good level of granularity and have rather homogeneous budgets and durations, as to define a cumulated S-curve of actual cost that may be reasonably approximated to a straight line.

Project managers have to consider some detailed aspects inherent with the definition of performance indices.

One is that during the first stages of a project, because of the little number of performed activities, the CI fluctuates and tends to stabilize by the time the project is 20% complete (Lipke 2005). Thus, the CI can be considered as a reliable source of performance information and future indication only from that date. Furthermore, the CI is likely to worsen from that point in time as the project progresses because of schedule delay, rescheduling and rework, which increase as more activities unfold (DoE 1980; Oberlander 1993).

Another is about the SI calculation metrics (Eq. 9.5b): as far as the project progresses the SI tends to get close to 1 even if the project is behind schedule. Indeed, at the finish date when the work performed (WP) equals the work scheduled (WS), the schedule variance is nil and the schedule index is 1. As a result, the SI and the associated formulae are useful until the project is no more than 70–80% complete (Fleming 1992).

In light of the consideration that growing expenditures involves declining control, it may be concluded that project monitoring is a valuable support to project control decisions and actions when the project is from 20 to approximately 80% complete, which is actually the most effective period to take project control actions. Earlier adjustments may rely on blind performance assessment, while later decisions may be ineffective, expensive and even get things worse.

9.5 Project Reporting

Project Reporting involves recording, editing and distributing documents containing information about budget, status and performance of several aspects of a project, such as scope, time, cost, cash flow, quality, safety, etc. Performance metrics typically are defined in preparation for project monitoring before project control. Reporting has to facilitate project communication and to enable effective project control processes at various organizations and management levels involved in the project development. To this end, it is very important that the project team activates a timely reporting scheme based on real-time measurement of performance.

The most useful monitoring documents report about time and cost performance. They are of two types, namely internal reports and contract reports.

Contractors use internal reports to monitor the project status with regard to cost, time and future trends. An internal periodical report (usually prepared on a monthly basis) is composed of a Cost Control Report and a Risks/Opportunities Report, which investigates the challenges that are likely to encounter in the next period. Typically, a monthly *Cost Control Report* describes the project status at the current date and contains information about:

- cost performance: budget cost, actual cost, earned value, cost variances and indexes;
- schedule performance and revised schedule;
- financial status: accounted cost, revenues, cash out, payments and cash in;
- estimates to completion.

Contract reports are basically used to monitor the financial status of the contract between the owner and contractor. It usually contains a schedule review, a schedule of value and the inherent certificate of payment, and a request for extra works, if a change order is applicable. In more details:

- the Revised Schedule contains a detail of progress report (individual activity, total progress), progress s-curves (actual, scheduled, forecast), resource-load profiles (actual, scheduled, forecast);

- the Schedule of Values is made of the Quantities Book, the Account Register, the Main Summary Account Register, and the Certificate of Payment of work done, which states the amount of money the contractor can charge (once deducted the value-retention guarantees and liabilities);
- if this is the case, a Change Order for extra scope or extra works includes a Request of Extra Works, a Report of extra works, and an Extra Work Schedule of Values with the inherent Certificate of Payment.

9.6 Areas of Project Control

Project control methods must be applied in conjunction with close and continuous monitoring in order to keep projects on track in terms of cost, quality and time. One control method may be characterized as performance-driven, using methods of project crashing or re-allocating resources to bring the project back on track. Another method is target-driven, suggesting that changes in the original contract may be necessary to re-align the project.

The basic process of project control is outlined below in Fig. 9.5.

Here discrepancies between initial targets and project targets are evaluated. The project may be executed and then controlled based on correcting the performance, or control actions may affect changes to the initial plans. We may call the first types of controlling actions as "performance-driven", while the latter go under the forms of "target-driven" corrective actions.

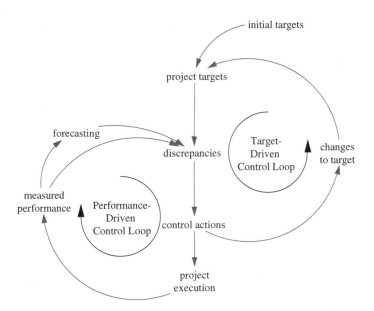

Fig. 9.5 The cyclical process of project control requires both perfromance-driven and target-driven corrective actions

Typically, *performance-driven control* is likely to take effect during the first part of project development when changes to resources and technologies based on project crashing may impact on productivity, schedule and performance.

Control adjustments may be required based on project performance for a variety of reasons. Perhaps a client has specified changes to the original plans. The final quality of the product delivered may differ from that originally agreed upon the contract. Oftentimes a technical challenge may arise, or original plans were made in error. Changes to the market may alter the economic feasibility of a project.

Whatever the reasons, measures of performance may lead to discrepancies with original contract documents. Constant monitoring and control of performance issues is vital to timely and costly completion of a project.

Figure 9.6 outlines this importance: corrective action taken at month 6 may be all it takes to bring the project within planned budget and time constraints. If corrective control action is delayed until month 10, it may be too late to adequately compensate for the problems. In this case, corrections made in month 10 lead to a project that is both over budget and delayed in schedule.

The ability to make corrective changes like the ones shown above is a reflection of a project's flexibility. Ideally, enough flexibility will be present to allow for changes that will solve problems in the three key areas of time, cost and quality. But the reality exists that performance-driven control may only be able to solve one of these issues at a time. Hence, project managers must be prepared to weigh the tradeoffs and priorities associated with time, cost and quality in control decisions. Only after a careful analysis of these tradeoffs should a control plan be executed.

Situations exist where the best possible plan of control action is based on the least harmful set of tradeoffs, as no good set of tradeoffs exists. Here resources must

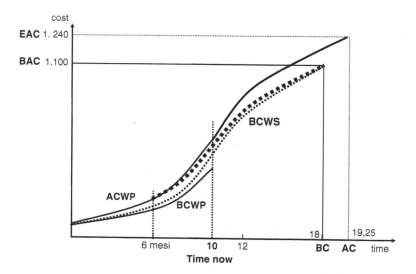

Fig. 9.6 Performance-driven corrrective actions are effective early in a project

be reallocated from non-critical activities to ones of a critical nature in a practice called "triage".

Another tool available to project managers is known as project crashing (see 7). This is the term given to the control method of adding either new resources or better methods of production in order to increase productivity. Perhaps this may lower the cost of non-critical activities or reduce the required time spent on critical activities. Consideration must be given to the economic tradeoffs when implementing project crashing, as it will likely come at an additional cost.

However, it often happens that late in the project, performance corrections are ineffective to reduce delays and cost overruns, so that a schedule and contract price revision is required. This target-driven control loop can be seen on the right side of Fig. 9.5.

Perhaps original budget or time estimates were unrealistic. In some cases, unforeseen changes to the market may make original cost or time deadlines impossible to meet.

This method of monitoring changes in schedule and making schedule adjustments accordingly is subject to contract compliance. If the main contract provisions do not allow for a schedule revision, cost of delays are likely to be paid by the party that is responsible for the schedule slippage. In order for the schedule adjustments to be taken, an understanding must be reached between the contractor performing the work and the owner. This will typically require some sort of modification to the original contract, commonly through a change order (Chap. 5).

References and Additional Resources About Project Monitoring & Control

Bennet J (1985) Construction project management. Butterworths, London
CII Construction Industry Institute (2004) Project control for construction. CII, Austin, TX, pp RS6–5
Christensen DS (1999) Using the earned value cost management report to evaluate the contractor's estimate at completion. Acquisition Rev Quart (Summer) 1999:283–295
Christensen DS, Antolini RC, McKinney JW (1995) A review of estimate at completion research. J Cost Anal Management (Spring) 1995:41–62
De Marco A, Briccarello D, Rafele C (2009) Cost and schedule monitoring of industrial building projects: case study. J Constr Eng Manage 9(1):853–862
Eldin NN (1989) Measurement of work progress: quantitative technique. J Constr Eng Manage 115(3):462–474
Fleming QW (1992) Cost/schedule control systems criteria: the management guide to C/SCSC, Rev edn. Probus Publishing Company, Chicago, IL
Hendrickson C (2008) Project management for construction, Online edn. http://www.ce.cmu.edu/pmbook/
Holm L, Schaufelberger J (2002) Management of construction projects. Prentice Hall, Upper Saddle River, NJ
Jung Y, Woo S (2004) Flexible work breakdown structure for integrated cost and schedule control. J Constr Eng Manage 130(5):616–625
Lipke W (2005) Connecting earned value to the schedule. Crosstalk: J Defense Software Eng 18(6)

Meredith JR, Mantel SJ Jr (2006) Project management: a managerial approach, 6th edn. Wiley, Hoboken, NJ

Oberlender GD (1993) Project management for engineering and construction. McGraw-Hill, New York

Pierce DR Jr (2004) Project scheduling and management for construction, 3rd edn. Reed Construction Data, Kingston, MA

Project Management Institute (2008a) A guide to the project management body of knowledge, 4th edn. Project Management Institute, Newtown Square, PA

Project Management Institute (2008b) A guide to earned value management. Project Management Institute, Newtown Square, PA

Ritz GJ (1994) Total construction project management. McGraw-Hill, New York, NY

U.S. Department of Energy (DoE) (1980) Cost & schedule control systems criteria for contract performance measurement. Implementation Guide. DoE, MA-0087, May 1980

Winch GM (2002) Managing construction projects. Blackwell Publishing, Oxford, UK

Woodward JF (1997) Construction project management getting it right first time. Thomas Telford, London

Yang YC, Park CJ, Kim JH, Kim JJ (2007) Management of daily progress in a construction project of multiple apartment buildings. J Constr Eng Manage 133(3):242–253

Part IV
Material Resources

Material resources, including construction materials and equipment, are fundamental to the development of a construction project from design through procurement to physical erection. Also, they represent a large portion of the total budget.

Materials management is the expertise that mainly copes with procurement management and site management. Primarily, procurement management (Chap. 10) has to ensure the flow of construction materials and the availability of specialty subcontractors to the construction sites. This is done through an integrated procurement process that involves planning, monitoring and control of the physical flow of material resources (design, manufacturing, transportation, and handling) and the administrative and contract implications towards suppliers and subcontractors.

Site management (Chap. 11) is involved with directing the construction site, the field logistics and building operations, with special focus on quality and safety. Typically, site management is responsible for equipment, acceptance and control of materials shipped to the site, storage and subcontracting management. Building activities require site managers to oversee construction through inspections, reviews, quality assurance procedures, organization and direction of foremen and workers.

Chapter 10
Procurement Management

10.1 Introduction to Procurement Management

Procurement is referred to as a set of activities designed and performed to assure regular flows of materials and services, according to a plan. In this broader sense, procurement includes the purchasing activities performed to secure an agreement between the buyer and the supplier, and the services that are needed for executing the project.

Procurement applies to all kind of supply, including construction materials and equipment, subcontract and professional services. The procurement process has to be carried throughout the project life-cycle: the owner may need to procure design, consulting and project management services in the early stages, as well as the construction delivery entity; material resources and specialty subcontractors are procured during execution; testing and maintenance services are finally necessary to finalize the project commissioning.

10.2 Procurement Methods and Strategies for Managing the Construction Supply

The definition of a proper strategic framework for material procurement enables a collaborative environment and successful relationships between owners, contractors and subcontractors (i.e. win-win environment).

First, the strategy has to be set at the organization level, where the procurement function is designed and its management responsibilities are assigned. In some construction organizations, a centralized procurement division leads the task (Fig. 10.1).

In this case, the various project teams place requests to the central purchasing department. In turn, this would keep available stocks of commodities and standard items to timely supply the construction sites and to obtain lower costs for bulk purchasing. The adoption of computer-based systems, such as material requirement planning (MRP), helps in the task of supply and stock management.

In other companies, the task is decentralized (Fig. 10.1) and assigned to the project teams, which may appoint dedicated procurement managers. In this

A. De Marco, *Project Management for Facility Constructions*,
DOI 10.1007/978-3-642-17092-8_10, © Springer-Verlag Berlin Heidelberg 2011

Fig. 10.1 Organizational models of the procurement unit. From *top* to *bottom*: centralized procurement, decentralized project-based procurement and mixed procurement approach

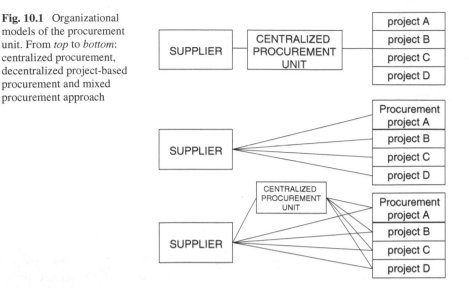

case-scenario, procurement is based on specific needs, takes advantage of local trade conditions, and reduces the cost of material stock and inventory management due to an effective just-in-time delivery of materials to the construction site.

To take advantage of both models, the procurement function may work based on a mixed approach (Fig. 10.1): bulk materials, commodities and standardized items are bought in large quantities from the low supplier by a centralized unit based on multiple-project material requirement estimates, while specific trade and pre-fabricated components are ordered based on specifications and schedule constraints by procurement managers as part of the project management team.

Procurement organizational units are responsible for searching and selecting vendors, for keeping relationships with suppliers and subcontractors, for expediting supplies and for managing and administrating the procurement process. A centralized procurement division may better consolidate long term know-how, expertise and methods required for the task with special regard to market and vendors' knowledge, and may reduce the cost of material supplies.

Project procurement units are suitable for non-standardized projects, where a partnering relationship with the supplier is required to fabricate special items and a strong on-site continuous relationship has to be established with construction subcontractors.

Second, the key decision in defining the procurement strategy for a construction project is whether "to make" or "to buy" (i.e.: a general contractor may decide to perform the electrical works either by buying electrical materials and using own craftsmen or by subcontracting the complete job). A "buy" strategy usually benefits from subcontractor productive specialization, variations in labour cost, flexibility in resource usage and on-site capacity. The more the production process is assigned to subcontractors, the more the need to establish partnering relationships in order not to jeopardize the project because of unfair subcontractors (Ritz 1994).

Indeed, a competitive client-vendor procurement model usually involves a large number of suppliers with a high turnover rate, nitpicking contracts, and uneven negotiations, which in turn may result in low quality, frozen stocks for rejection of materials, and unsteady material inflows. In contrast, the establishment of partnerships with suppliers creates conditions for cooperation and synergies aimed at co-design and co-making, supports decrease of stocks and procurement costs due to a lower number of suppliers, standardization of procedures and a continuous increase in quality. As a drawback, the contractor may become too much dependent from the supplier and decrease the competitive effort for innovation.

Thus, the activity of selecting and rating reliable suppliers is of crucial importance also because more and more bidders compel contractors to list their selected subcontractors and suppliers in the bid (Fisk 2003). Vendor rating is aimed at advantaging suppliers with best performances, rationalizing and quantifying qualitative aspects of procurement, making clear and collaborative the relations, as well as monitoring changes in the suppliers' behavior and in the market conditions over time.

Material suppliers and construction subcontractors are usually rated based on product or service quality, technological level, price, and financial reliability. International quality certification standards can help in the first task of grading suppliers, such as the norms ISO 9000–2001.

Then, more in-depth metrics and indices may be used to rate vendors and subcontractors. For example, the quality vendor rating is the output of the summation of the accuracy index (number of shipments compliant with the initial order) and of the product testing index (number of items passing the quality check); the service vendor rating is the outcome of the punctuality index (goods shipped on time or activities completed as scheduled), order processing lead time index, flexibility index (promptness in making changes to the original order).

10.3 The Procurement Process

The procurement process applies to all classes of materials with little variation. The sequenced activities of a procurement process of fabricated items are usually the following.

- A *Request for Purchase* is usually advanced from the engineering team or from the construction site to the procurement division.
- Based on the pieces of information, technical specifications, material take-offs (MTO) and drawings included in the request for purchase, the procurement team submits a *Request for Bid* to the qualified and rated suppliers or trade subcontractors. For large-scaled procurements, there may be a *Pre-Selection* of possible specialized traders.

- Once the bids are received, the *Bid Analysis* is performed by engineering and procurement personnel, to ensure that both technical and commercial points are covered.
- Out of the analysis, the *Selection* is made directly or after a *Negotiation* between the low bidders.
- The purchase process is completed with placing the *Order* or, if applicable, a complete *Contract* including general conditions (applicable to all orders from the same entity), special conditions (particular conditions that apply only to the individual order), contract clauses and technical specifications (specs).
- Finally, the process requires operative and administrative *Monitoring and Control* from the date the order is issued to the final supply or completion of subcontracting on-site activities.

Figure 10.2 is the flow chart for the purchase process of bulk materials.

Once the order has been issued and confirmed by the supplier, the monitoring and control phase of material procurement has to be carried out.

This mainly copes with supply expediting processes, which require monitoring and control activities executed at the manufacturer facility to obtain accurate and real-time information about the progress status of the supplies and to assure the respect of quality and production lead-times. In this process, testing is of great importance: acceptance of materials is subject to documentation, certificates, intermediate and final tests performed before shipping of materials to the construction site.

Logistics and transportation are also an inherent part of the material control process. A transportation plan includes analysis about the better types of transportation contracts that can be used, the transportation means and routes with regard to

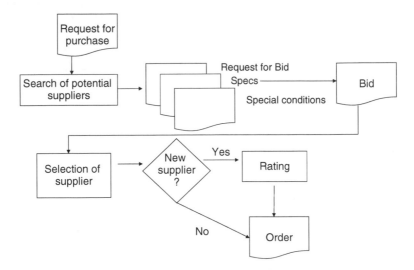

Fig. 10.2 Flow chart of bulk materials procurement process

Table 10.1 Types of transportation agreement

Code	Description	Mean of transport	Customs formalities by	
			Seller	Buyer
EXW	Ex works	All		EX IM
FCA	Free carrier	All	Export	Import
FAS	Free alongside ship	Sea		EX IM
FOB	Free on board	Sea	EX	IM
CFR	Cost and freight	Sea	EX	IM
CIF	Cost, insurance and freight	Sea	EX	IM
CPT	Carriage paid to	All	EX	IM
CIP	Carriage and insurance paid to	All	EX	IM
DAF	Delivered at the frontier	All	EX	IM
DES	Delivered ex ship	Sea	EX	IM
DDU	Delivered duty unpaid	All	EX	IM
DEQ	Delivered ex quay	Sea	EX IM	
DDP	Delivered duty paid	All	EX IM	

quantities and timing of freight delivery. Transportation preliminaries also involve travel permits and documents that need to be filled in compliancy with border and authorities requirements.

Often, in the case of major items, transportation contracts have to be signed. There are different modes of transportation (classified by INCOTERMS 1990), including custom clearance procedures and kinds of insurance coverage.

Table 10.1 is a list of possible transportation contract arrangements between the supplier and the purchaser.

References and Additional Resources About Procurement Management

Clough RH, Sears GA (1994) Construction contracting, 6th edn. Wiley, New York, NY

Fisk ER (2003) Construction project administration, 7th edn. Pearson Education, Upper Saddle River, NJ

Hendrickson C (2008) Project management for construction, 2nd edn. Carnegie Mellon University, Pittsburgh, PA. http://www.ce.cmu.edu/pmbook/

Ritz GJ (1994) Total construction project management. McGraw-Hill, New York, NY

Chapter 11
Site Management

11.1 Management of Construction Equipment

Field equipment is a necessary resource to support the construction effort. This includes site temporary facilities, building tools, and construction machines used for the purpose of physical execution of the project, such as excavating, loading, compacting, drilling, erecting, mixing, etc.

The cost of purchasing, usage and maintenance of construction equipment can be considered as a project overhead cost. Thus, project teams have to look for the most effective tradeoffs between time and cost of equipment utilization. Resource-based scheduling of construction equipment usage is recommended as part of the project schedule. As well as all the other human and material resources, equipment has to be consumed on a timely manner depending on the construction schedule. Sometimes adjustments to the schedule and added dependences to the sequencing of construction activities may help in levelling the workload of equipment and save money from peak needs of extra machines (Hendrickson 2008).

Also, site managers often have to make decisions about purchasing or renting the construction equipment. Purchasing involves depreciation or amortization if there is a supporting financial institution, maintenance and failure charges, while renting may be more expensive on the short term. Decision analysis tools (see Chap. 12) may help in the task of decision making.

Depreciation is defined to as a cost that reduces the initial value of an asset as a result of its obsolescence. In other words, the initial investment is prorated over the aging timeframe (e.g. 5 years) and the cost per hour of the machine usage can be obtained. In order not to make difference between new and old assets, it is required that standardized mean depreciation cost index are defined for each category of construction equipment.

To avoid the complexity of such a system of equipment cost monitoring, some general contractors prefer to rent the equipment from a dealer, which may be a commercial trader or a parent company-owned entity. The advantages of using a child company that leases the equipment to the projects are the following:

- all construction sites are invoiced the actual usage of equipment thus facilitating project cost monitoring;

- economic results are indifferent to the parent company;
- preventive maintenance and safety are enhanced thanks to the centralized technical and quality control;
- underutilization of equipment is avoided since machines can be rented out to external customers during the idle period; therefore, the creation of an equipment entity within the firm allows for transforming the site equipment from a cost into a partial revenue.

11.2 Quality Management

Quality is an inherent aspect of construction and involves the physical implementation of a project (materials, supplies, activities, etc.), as well as the operational environment of the project with regard to methods, processes, tools, and people.

Quality is usually referred to as the technical specification of the scope of work: in most traditional contracts quality is a constraint and the contractor is responsible for the delivery to the owner of the specified level of quality in terms of construction materials, finishing, equipment, and resources used in the building process. When design and construction phases are overlapped, such as in CM, DBB and turn-key contracts, the level of quality has to be specified during the design development and can be adjusted while construction is underway within the limits and requirements specified in the contract.

To successfully implement the contract, all the parties are required to control and assure that the project precisely respect the quality specifications all along the project development. Stakeholders, owners, construction managers and contractors are involved in the process of quality assurance and the role may be delegated to several inspectors.

Local building departments, safety and health inspectors, insurance company inspectors and financial institution inspectors make it possible for the stakeholders to keep an eye on the project. The Resident Project Representatives and Architect's superintendents control quality on the behalf of the owner. Finally, the project management is the operational entity committed to deliver and assure quality by the effective action of site managers, quality managers, and foremen.

Quality control and quality assurance are some of the main procedures available to the project management team.

Quality control procedures are usually agreed upon the contract. Contracts require validation-passing gates such as quality checks, testing, walkthroughs, and inspections. In such circumstances, the parties are placed in adversarial positions: at the end of major phases during the production the inspectors appointed by the owner control that the contractor provides the level of quality specified in the contract, while the contractor's managers do the same control on subcontractors and material vendors.

Quality checks and performance tests are the elementary steps in helping to ensure that site operations and materials respect the specified level of quality:

- Checking construction workmanship is of great importance, in the sense that both the output and the process of doing it have to be controlled: the quality of site production can be checked through technical testing, while performance of execution may be measured with human resources productivity, which is most often tightly linked to the quality of execution.
- As far as construction materials are concerned, either the contract specifications or the "best-available grade" define the minimum level of their required quality. Quality assurance for materials and equipment must be established to assure the satisfactory performance. Site inspectors and managers are responsible for accepting quality-checked materials and may reject or ask for substitution of faulty materials.

A walkthrough is a semi-formal work quality control task. The purpose of the walkthrough is to notify the stakeholders (e.g.: the owner) that a portion of the scope of work is complete, and get approval. Typically, gates associated with minor milestones or specific work units are passed by means of a walkthrough.

An inspection is a formal review of the quality of a portion of the work done, as well as of the quality of the process itself. A typical inspection is the substantial completion, as a pre-finish validation gate allowing for commissioning the constructed facility for occupancy, even if minor final works still need to be finalized. The substantial completion inspection is a multi-stage, formal process with legal and contractual significance. An inspection is usually aimed at checking out a punch-list of open issues and see when those aspects will be completed.

The drawback of validation-passing gate approaches is that usually the production people tend to cover and hide their mistakes to the controller/managers by nature. The pressure on quality by the control team often causes more hidden errors and results in lower quality. In turn, quality control inspection delays the detection of cumulative errors which require more time and money to recover or rework.

Yet, quality is not only a primary objective (together with cost and time) and contract-specified outcome in a constructed facility, but also a general philosophy by which process are carried in a Total Quality Management (TQM) perspective.

A TQM approach requires that operational methods – such as validation-passing gates and quality assurance procedures for activities and supply of materials – are executed under a more general framework aimed at continuous improvement of the organization and personal growth of its individual members. Quality is viewed in the broadest sense including the well-being and satisfaction of all people involved and the creation of long-lasting relationships with clients and suppliers (MIT Open Courseware).

11.3 Site Operations and Safety

Site operations represent the core activities of a construction project where the larger part of the capital investment is spent in the physical erection of the facility. Thus, it is important that the highly complex and various jobsite operations are efficiently and effectively managed: the site managers have to work close to the project management to put the project into action.

This involves the continuous application of project management practices to a detailed scale in several areas:

- planning and scheduling (keeping track and organize the personnel on the jobsite; preparation and implementation of weekly -and sometimes even daily- schedules with detailed plans for crews; etc.)
- monitoring and control (progress monitoring, book-keeping, performance driver measurement, problem solving and corrective actions; etc.);
- physical implementation (preparation of detailed specs, drawings and erection procedures to be used by foremen and workers on the site; technical instruction and directions; etc.);
- communication (project reporting; client and subcontractors communication management; human resources management; etc.);
- logistics and materials management;
- regulations and permits.

Also, since site operations are inherently dangerous to design and control a safety plan. The safe execution of construction activities and the safe usage of site equipment must be a major concern to site managers to protect human resources from injuries and severe damages, to comply with construction safety regulations issued by specific national agencies (such as OSHA in the U.S.), and to avoid the extra cost and time in case of accidents. A safety plan usually analyzes the possible sources of risk on the jobsite and prescribes measures, procedures, systems and actions that may prevent accidents and safeguard personnel.

A safety plan must contemplate the cost for its implementation, including the salary of a safety manager.

References and Additional Resources About Site Management

Fisk ER (2003) Construction project administration, 7th edn. Pearson Education, Upper Saddle River, NJ
Hendrickson C (2008) Project management for construction, 2nd edn. Carnegie Mellon University, Pittsburgh, PA. http://www.ce.cmu.edu/pmbook/
Ritz GJ (1994) Total construction project management. McGraw-Hill, New York, NY

Part V
Uncertainty

So far, all management methods and techniques presented along the book assumed certainty about the future outcomes of a project. As a matter of fact, much uncertainty exists in project management with regard to a variety of issues, so that it is hard to carry out many tasks, such as budgeting and scheduling, solely with deterministic approaches. Other more complex methods are necessary to take uncertainty and risks into consideration.

One method is to make decision analyses based on multiple scenarios and simulation under uncertain conditions (Chap. 12). Another is to use probabilistic scheduling techniques such as PERT (Chap. 13). Finally, project risk management methodology is suggested as a way to bring all aspects of project variations and foreseeable uncertainty under the control of the project team (Chap. 14).

Chapter 12
Decision Making

12.1 Introduction

Risk refers to as uncertainty about consequences that individual and organizations face while putting plans into action.

Construction project management has to cope with several kinds of risk, such as weather conditions, different productivity than expected, defective work by contractors and subcontractors, financial instability, procurement lead times, lawsuits, labor difficulties, unexpected manufacturing costs, failure to find sufficient tenants, community opposition, unrealistically low bids, late-stage design changes, unexpected subsurface conditions, permit and authorization obstacles, etc.

Risk in construction projects has tremendous impacts and much time in construction management is spent focusing on risks, even if still companies do not have conscience of their huge management effort dedicated to this task and have little formal methods to support it. Many practices in construction are driven by risk; bonding, insurance, licensing, and the definition of an appropriate contract structure are some examples of this.

Risks may cause different effects depending on the decisions that project teams make: consequences are the outcome of decision making.

12.2 Decision Analysis

Suppose a commercial real estate developer and construction company wants to construct a residential apartment building. As one of several design options, the design team is wondering either to construct the roof with metal plates (option A) or pre-cast concrete ones (option B).

The estimated cost for option A is 1 million dollars. Since the steel usually has important cost fluctuations, estimators are envisaging three possible scenarios for the cost of steel at the time the construction will take place: (1) 25% is the likelihood for the steel cost to remain steady, (2) 25% is the likelihood for the steel cost to decrease by 30%, and (3) 50% is the probability that the steel cost may increase by 30%.

With option B, the estimated cost is 800,000 dollars, with (1) 10% likelihood for the cost to stay firm, (2) 20% to go down by 10%, and (3) 70% to go up by 40%.

Given these estimates, and – for a simpler reasoning – assuming that the choice is not affecting other construction elements and cost, how can the project team evaluate the different options?

12.2.1 Decision Trees

The first step of decision analysis is to identify the different alternatives, associated risks and the outcomes. In doing this, decision tree analysis is useful and viable formal technique. Several commercial software application exist to assist in the task, such as Tree Age®, Vanguard®, and many others.

Decision trees may be used both for illustrating decision making with uncertainty and for quantitative reasoning. They allow for representing decisions through the flow of time, uncertainties via events, and consequences that may have deterministic or stochastic behavior. The decision tree is a network diagram that illustrates the sequence of decisions and events with the associated chance of occurrence (Haimes 2004).

The decision tree of this case-project can be constructed as in Fig. 12.1.

The box node in the tree indicates a Decision Node: whether to build a metal or a concrete roof. The circle node indicates a Chance Node which is the stage where possible events occur. The triangle node at the right-side end is a Terminal Node, and this indicates the outcome associated with the previously made decisions and events realized.

In this problem thus there are two branches from the Decision Node in the figure. With regard to possible events, there are three scenarios of possible future cost: unchanged cost, decreased cost, and increased cost. Therefore, the Chance Nodes after each decision branch have three branches of events with their known associated

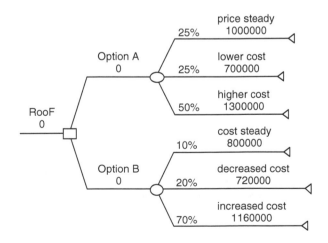

Fig. 12.1 Decision tree of the residential building roof

probabilities. The decision tree is then complete with calculation of the outcome of each branch.

In this case, we did not consider the different construction task durations of the two alternatives. Usually, the outcomes can be better calculated by computing the Net Present Value (NPV) for each scenario to take the money value of time into account.

12.2.2 Expected Value

Now with the complete decision tree on hand, the project team can decide which plan is better for the company. The most common and easy way to do this will be to calculate the expected value of each decision strategy and pick the decision that maximizes this value.

The expected value (E) is computed by summing the products of each outcome by the associated chance.

Thus, the expected values for option A (metal plates) and plan B (concrete slabs) are:

$$E \text{ (Plan A)} = 1,000*0.25 + 700*0.25 + 1,300*0.50 = 1,075 \text{ (thousand dollars)}$$

$$E \text{ (Plan B)} = 800*0.1 + 720*0.2 + 1,160*0.7 = 1,036 \text{ (thousand dollars)}$$

Given the probabilities of possible events, we could calculate the expected values, and, though the difference is small, Option B turns out to be less expensive than Option A.

Yet, in real world projects, probabilities of events are not always available, such as in the case of variation of construction material market prices: it is likely that probabilities of events are empirically estimated. Thus, it may be helpful to use other decision rules that do not rely on expected likelihood of braches, but just look at possible worst and best-case outcomes.

Since risk preference may depend on risk attitude: risk-loving individuals and organizations may look for the maximum benefit disregarding the little related probability of occurrence. Instead, a risk-averse decision maker will pick the option that limits losses or maximizes minimum gain.

With reference to Fig. 12.1, risk lovers seeking for maximum benefit will pick the "optimistic" Option A of building a metal roof, which may lead to the lowest possible cost (700,000 dollars).

Under a "maximin" rule or, more properly in our case, the "minimax" one, a risk averter would choose Option B that minimizes the maximum loss (1,116,000 dollars worst-case cost for Option B is less than 1,300,000 in A). This means that the project team would protect the company against some worst case cost.

Other and more complex rules may be applied, such as the "regret" criterion: the benefit actually received and the maximum benefit that could have been obtained if the appropriate choice had been made (de Neufville 1990).

12.2.3 Utility Function

From the above examples, decision rules depend highly on whether individuals or organizations are optimistic or pessimistic. The attitude of an organization to risk sometimes affects the decision even when we know the exact probabilities of possible events. People are not indifferent to uncertainty and the lack of indifference from uncertainty arises from uneven preferences for different outcomes (Bedford and Cooke 2001); for example, someone may dislike losing money far more than gaining it or value gaining money far more than they disvalue losing it. Individuals differ in comfort with uncertainty based on circumstances and preferences and risk-averse individuals will pay for "risk premiums" to another party to avoid uncertainty. Risk attitude is a general way of classifying risk preferences: risk averter fear loss and seek sureness, risk neutral are indifferent to uncertainty, and risk lovers hope to "win big" and don't mind losing as much. Also, risk attitudes change over time and circumstance (MIT Open Courseware).

Consider the roofing case again. Given the probabilities of materials cost growth or reduction, we conclude that Plan B is better than Plan A based on the calculation of the expected values of both decisions. However, this conclusion may change if the company has a risk-lover attitude.

Suppose for the time being that the company is rather a risk-lover, and has the "utility function" shown in the chart below (Fig. 12.2).

Utility theory states that individuals look for maximization of their expected utility out of an uncertain outcome. The expected utility is a measure of the individual's implicit preference, for each case-scenario in the risk environment. It is represented

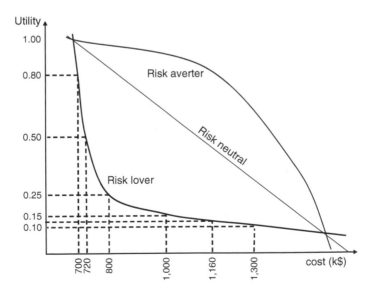

Fig. 12.2 Utility function

by a numerical value associated with each monetary gain or loss in order to indicate the utility of these outcomes to the decision-maker (Flanagan and Norman 1993, Chap. 5).

Now, the Expected Utility Value of Option A and B are calculated using this Utility Function, which multiplies the utility value of an outcome by its probability.

$$\text{EUV (Option A)} = \text{Utility Value } (1{,}000\,\text{k\$ })*0.25 + \text{UV}(700\,\text{k\$ })*0.25$$

$$+ \text{UV}(1{,}300\,\text{k\$ })*0.50$$

$$= 0.15*0.25 + 0.8*0.25 + 0.10*0.50 = 0.2875$$

$$\text{EUV (Option B)} = \text{UV}(800\,\text{k\$ })*0.1 + \text{UV}(720\,\text{k\$ })*0.2 + \text{UV}(1{,}160\,\text{k\$ })*0.7$$

$$= 0.25*0.1 + 0.50*0.2 + 0.125*0.7 = 0.2125$$

The calculation of Expected Utility Value indicates that Plan A would be better than Plan B if we had a risk-lover attitude. Once the utility function of an organization is established, decisions can be made based on the calculation of Expected Utility Value in the same way as in this example.

12.2.4 Notion of Risk Premium

The reasoning so far can be useful to better understand the notion of risk premiums in construction and, particularly, the risk sharing policies that apply to a construction contract between the owner and the contractor (Chap. 4).

A risk premium is the amount paid by a risk-averse individual or organization to avoid risk. It is very common that an owner pays higher fees to reputable contractors and higher charges by contractor for risky work or for bearing the financial risk of a turnkey contract.

Thus, in a fixed-price or turnkey contract, the price is higher than the one that could be estimated based on time & material because it includes a contingency paid to the contractor as a risk premium for shouldering risk of extra unforeseeable cost. The owner is the risk-averter party and the contractor is the risk lover that takes advantage of future benefit or disadvantage of future loss.

To calculate the value of the risk premium that an owner may be willing to pay a turn-key contractor it is useful to introduce the notion of "Certainty Equivalent" (Lifson and Shaifer 1982).

To this end, consider a risk-averse owner with the preference function f in Fig. 12.3 faced with a contract that may provide:

- 50% chance of saving +$20,000 (the positive sign indicates a gain)
- 50% chance of cost –$10,000 of cost overrun (the negative sign indicates loss)

The average money that can be saved in the project equals:

Fig. 12.3 Utility function for
cost saving

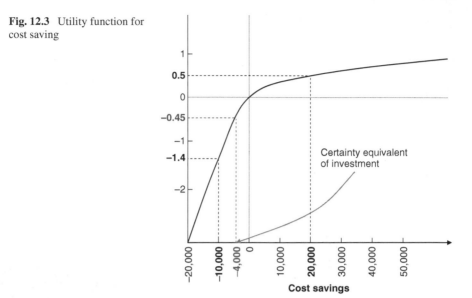

Expected value $= 0.5*\$\,20{,}000 + 0.5*\$ - 10{,}000 = \$\,5{,}000$

and the average satisfaction with the investment is the following:

Expected utility value $= 0.5*f(\$\,20{,}000) + 0.5*f(\$ - 10000)$

$$= 0.5*0.5 + 0.5* - 1.4 = 0.25 - 0.7 = -0.45$$

This owner would then be willing to pay for a sure risk premium yielding satisfaction greater than –0.45, which means that she can get –0.45 satisfaction for a sure:

$$f^{-1}(-0.45) = -\$\,4{,}000$$

We call this the certainty equivalent to the contract: the owner should be willing to pay a sure fixed-price that includes an additional $4,000 risk premium to the contractor shouldering all possible cost overruns. The owner would be willing to trade the uncertain project for any certain risk premium that is less than $4,000.

More generally, consider the situation in which there is uncertainty with respect to a consequence c and a non-linear preference function f.

Assume that:

- $E[c]$ is the mean outcome of the uncertain investment c (in the previous example, this was $0.5 * \$20{,}000 + 0.5 * \$10{,}000 = \$5{,}000$)
- $E[f(c)]$ is the mean satisfaction with the investment c (in the example, this was $0.5 * f(\$20{,}000) + 0.5 * f(\$-10000) = 0.5 * 0.5 + 0.5 * -1.4 = 0.25 - 0.7 = -0.45$)

We call $f^{-1}(E[f(c)])$ the certainty equivalent of c, which represents the size of sure return that would give the same liking as investment c (in the example, this was $f^{-1}(-0.45) = -\$4,000$).

The shapes of the preference functions allow classifying risk attitude by comparing the certainty equivalent and the expected value:

- for risk taker individuals, $f^{-1}(E[f(c)]) > E[c]$, as they want certainty equivalent > mean outcome,
- for risk neutral individuals, $f^{-1}(E[f(c)]) = E[c]$,
- for risk averse individuals, $f^{-1}(E[f(c)]) < E[c]$, as in the example.

Consider a risk averse individual A for whom $f^{-1}(E[f(c)]) < E[c]$ and a less risk averse party B. A can lessen the effects of risk by paying a risk premium r of up to $E[c] - f^{-1}(E[f(c)])$ to B in return for a guarantee of $E[c]$ income. The risk premium shifts the risk to B while te net investment gain for A is $E[c] - r$, but A is more satisfied because $E[c] - r > f^{-1}(E[f(c)])$, so that B gets an average monetary gain of r.

12.3 Multiple-Attribute Decision Making

Frequently it is necessary to care about multiple attributes. For example, it is opportune to make decisions based on cost, time, and quality. The terminal nodes on decision trees can capture these factors, but still need to make different attributes comparable (Haimes 2004).

Consider the following example. Decision has to be made about a waste water treatment station whether to replace, repair or keep it as is. Of course, it is of great importance to the decision maker to maximize the duration of the facility and minimize its cost.

The problem can be described using the decision tree in Fig. 12.4.

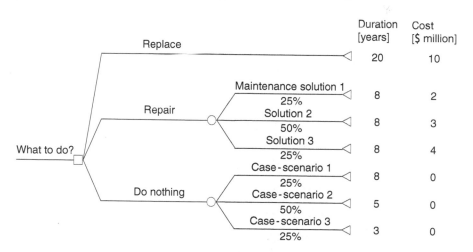

Fig. 12.4 Decision tree for the waste water treatment station

In Fig. 12.4, the replace option has the maximum investment resulting in the longest mean duration of the facility. Three are the possible case-scenarios for cost of maintenance. The do nothing zero-cost option has a 25% chance to get the same 8 years duration that can be obtained with maintenance, but has far more chance to lead to a failure of the facility in a few years.

Since the choice has to be made based on two parameters (time and cost), it is needed to compute the Pareto-optimal set for the decision.

First, we calculate the mean values of both attributes for each chance node:

- replace mean value = [10 years, $10 million]
- repair mean value = [8 years, 0.25 * 2 + 0.5 * 3 + 0.25 * 4] = [8 years, $3 million]
- do nothing mean value = [0.25 * 8 + 0.5 * 5 + 0.25 * 3, $0] = [5.25 years, $0]

Then, we plot a time/cost graph as in Fig. 12.5.

In this graph, the three option of the decision tree are all Pareto-optimal solutions, as well as any state of the problem on the Pareto-optimal curve above. Situation A, on the contrary, is dominated by the other options because it gives only 7 years for an investment of 5 million dollars: all situations below the Pareto-optimal curve are non-optimal because they are "dominated" by more efficient solutions and they are also referred to as inferior solutions.

Moreover, if a situation similar to B may exist, B would be an efficient solution (where the optimal curve is plotted from Do nothing through B to Replace) and Repair would result in a dominated situation. A decision is "Pareto optimal" (or efficient solution) if it is not dominated by any other decision.

Using the notion of Pareto-optimality, even if one cannot directly weigh one attribute versus another, one can rank some consequences and rule out decisions that are inferior with respect to all attributes. The key concept here is that one may not be able to identify the best decision, but we can rule out the obviously bad.

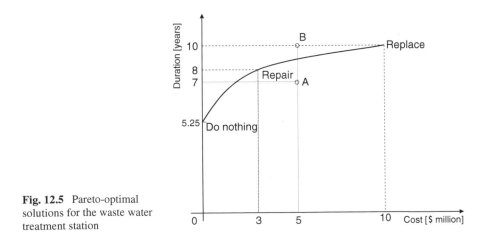

Fig. 12.5 Pareto-optimal solutions for the waste water treatment station

12.4 Monte Carlo Simulation

So far, we have considered a decision making problem with deterministic probabilities. However, no single phenomenon in the real world can be forecasted with an exact probability. Any events in the future will happen with some distribution of possibility which we may know from empirical data, if those are available and can be considered as reliable.

Computer-based Monte Carlo simulations are a helpful tool to evaluate decisions related to future events that may be described with probabilistic distributions. By random selection of a number of possible case-scenarios, this method allows computing the resulting distribution of project outcome in the terminal node. Then, comparing the distribution of terminal payoff of different options allows making the decision.

Of course, Monte Carlo simulation is of great value if the company has a large set of historical data from past projects and this can be obtained through the application of a systematic knowledge feedback process from all projects.

References and Additional Resources About Decision Making

Bedford T, Cooke R (2001) Probabilistic risk analysis: foundations and methods. Cambridge University Press, Cambridge
Chapman CB (2002) Managing project risk uncertainty: a constructively simple approach to decision making. Wiley, Hoboken, NJ
de Neufville R (1990) Applied systems analysis. McGraw-Hill, New York, NY
Flanagan R, Norman G (1993) Risk management and construction. Wiley-Blackwell, Hoboken, NJ
Haimes YY (2004) Risk modeling, management, and assessment. Wiley, Hoboken, NJ
Lifson MW, Shaifer EF (1982) Decision and risk analysis for construction management. Wiley, Hoboken, NJ

Chapter 13
Probabilistic Scheduling

13.1 Scheduling with Uncertainty

So far, scheduling has been presented as a deterministic method: the critical path network assumes that activities have predetermined duration and that these durations are known from the planning phase. This simple approach means that project managers are ignoring uncertainty. But because of either inexperience or different productivity in performing a given activity or risk and unforeseen events, the original schedule prepared using the CPM is likely to be vanished by the actual project performance (see Chap. 8).

Thus, managers have to improve their schedules by taking into consideration uncertainty.

An informal method is to apply time buffers. The project manager could schedule 6 weeks of work – using a deterministic scheduling method like CPM -followed by a 1-week buffer of no work. The manager may simply assume that whatever amount of time was lost in the first 6 weeks can be made up during the buffer week. Therefore, the project will stay on schedule. This method does add some flexibility to the project, but a series of "bad-luck" events could create delays that the buffers cannot cover (Goldratt 1997).

The first two options, ignoring uncertainty and adding time buffers, depend upon the magnitude of the project and also the experience of the project manager. A third set of methods, including "What-If Scenario analysis" and Project Risk Management practices (Chap. 14) is the only informal one in which a project manager will acquire any useful information regarding risk and uncertainty.

Formal probabilistic approaches are also available to project managers, as presented in this chapter.

13.2 Pert

PERT scheduling approach is closely related to a CPM scheduling: the difference is that it uses probabilistic estimates for activity durations. The result of the PERT network solution is a critical path, which has no more a deterministic length, but

A. De Marco, *Project Management for Facility Constructions*,
DOI 10.1007/978-3-642-17092-8_13, © Springer-Verlag Berlin Heidelberg 2011

rather an expected duration associated with its inherent probability, resulting from statistical issues.

For each activity into the WBS, three possible durations are estimated, possibly based on a database of historical data from past similar projects. These three estimations are referred to as the optimistic scenario (a), the most likely scenario (m), and the pessimistic scenario (b). Once determined, the three values can be used to deduce a Beta distribution. For a certain activity, the optimistic scenario represents the shortest possible duration, the most likely scenario represents the most common duration, and the pessimistic scenario represents the longest possible duration. For all three scenarios, the project manager must choose the values as accurately as possible. This is why the experience of the project manager is a major factor in using PERT to analyze project uncertainty.

Figure 13.1 represents a possible Beta distribution for a specified activity. The x-axis is time. Both the optimistic and pessimistic durations represent the length of time for the activity such that the project manager can be 99% sure that the actual duration of the project will be between values a and b.

With the values of a, m, and b, the expected duration, variance, and standard deviation can be calculated.

The expected duration, d, is

$$d = [2m + (a + b)/2]/3 = (a + 4m + b)/6 \qquad (13.1)$$

The standard deviation of the distribution, S, is a function of the optimistic and pessimistic values only. Since these values represent the 99% range of the Beta distribution there are exactly six deviations between a and b:

$$S = (b - a)/6 \qquad (13.2)$$

In fact, a and b are estimated at one-in-a-hundred level so that:

$$b - a = 2*3S = 6 S; S = range/6$$

The variance of the deviation, V, is simply the square of the deviation.

$$V = S^2 = [(b - a)/6]^2 \qquad (13.3)$$

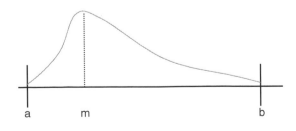

Fig. 13.1 Beta distribution of durations

Example – Computing the Expected Duration

A main contractor for the construction of a warehouse has hired a subcontractor to erect the pre-casted structures. The main contractor needs to know the duration of this event so it can plan linked activities accordingly.

To estimate the duration, the company has asked the subcontractor to estimate an optimistic, most-likely, and pessimistic duration for the task, as follows:

optimistic duration $a = 5$ days
most likely duration $m = 6$ days
pessimistic duration $b = 9$ days

Using Eq. (13.1), the expected duration is:
$d = (a + 4m + b)/6 = (5 + 4 \cdot 6 + 9)/6 = 6.33$ days
The standard deviation is calculated using Eq. (13.2):

$S = (b - a)/6 = (9 - 5)/6 = 0.67$

which makes the variance equal $V = 0.44$ from Eq. (13.3).

The Normal Distribution Table, where $V = 0.44$ has a corresponding likelihood value of 67%, helps us presume that the duration of 6.33 days is likely to occur with 67% probability.

The duration distribution examined in the example is skewed to the left, but distributions may be skewed to the right or be symmetric – normal distribution – as in Fig. 13.2.

The true significance of the PERT analysis is to determine the probability that an activity, or sequence of activities, will be completed in a certain amount of time.

The steps in PERT analysis are:

- obtain a, m and b for each activity of the network,
- compute expected activity duration and activity variance, then
- compute expected project duration using standard CPM algorithm,
- compute project variance $V = S^2$ as sum of critical path activity variance,
- finally, calculate the probability of completing the project using the normal distribution Table 13.1, which means assuming that the project duration is normally distributed.

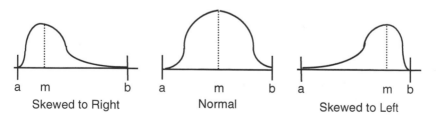

Fig. 13.2 Possible beta distributions

Table 13.1 Normal distribution chart

z	0.00	0.01	0.02	0.03	0.04	0.05	0.06	0.07	0.08	0.09
0.0	0.5000	0.5040	0.5080	0.5120	0.5159	0.5199	0.5239	0.5279	0.5319	0.5359
0.1	0.5398	0.5438	0.5478	0.5517	0.5557	0.5596	0.5636	0.5675	0.5714	0.5753
0.2	0.5793	0.5832	0.5871	0.5910	0.5948	0.5987	0.6026	0.6064	0.6103	0.6141
0.3	0.6179	0.6217	0.6255	0.6293	0.6331	0.6368	0.6406	0.6443	0.6480	0.6517
0.4	0.6554	0.6591	0.6628	0.6664	0.6700	0.6736	0.6772	0.6808	0.6844	0.6879
0.5	0.6915	0.6950	0.6985	0.7019	0.7054	0.7088	0.7123	0.7157	0.7190	0.7224
0.6	0.7257	0.7291	0.7324	0.7357	0.7389	0.7422	0.7454	0.7486	0.7517	0.7549
0.7	0.7580	0.7611	0.7642	0.7673	0.7704	0.7734	0.7764	0.7794	0.7823	0.7854
0.8	0.7881	0.7910	0.7939	0.7967	0.7995	0.8023	0.8051	0.8078	0.8106	0.8133
0.9	0.8159	0.8186	0.8212	0.8238	0.8264	0.8289	0.8315	0.8340	0.8365	0.8389
1.0	0.8413	0.8438	0.8461	0.8485	0.8508	0.8531	0.8554	0.8577	0.8599	0.8621
1.1	0.8643	0.8665	0.8686	0.8708	0.8729	0.8749	0.8770	0.8790	0.8804	0.8830
1.2	0.8849	0.8869	0.8888	0.8907	0.8925	0.8944	0.8962	0.8980	0.8997	0.9015
1.3	0.9032	0.9049	0.9066	0.9082	0.9099	0.9115	0.9131	0.9147	0.9162	0.9177
1.4	0.9192	0.9207	0.9222	0.9236	0.9251	0.9265	0.9279	0.9292	0.9306	0.9319
1.5	0.9332	0.9345	0.9357	0.9370	0.9382	0.9394	0.9406	0.9418	0.9429	0.9441
1.6	0.9452	0.9463	0.9474	0.9484	0.9495	0.9505	0.9515	0.9525	0.9535	0.9545
1.7	0.9554	0.9564	0.9573	0.9582	0.9591	0.9599	0.9608	0.9616	0.9625	0.9633
1.8	0.9641	0.9649	0.9656	0.9664	0.9671	0.9678	0.9686	0.9693	0.9699	0.9706
1.9	0.9713	0.9719	0.9726	0.9732	0.9738	0.9744	0.9750	0.9756	0.9761	0.9767
2.0	0.9773	0.9778	0.9783	0.9788	0.9793	0.9798	0.9803	0.9808	0.9812	0.9817
2.1	0.9821	0.9826	0.9830	0.9834	0.9838	0.9842	0.9846	0.9850	0.9854	0.9857
2.2	0.9861	0.9865	0.9868	0.9871	0.9874	0.9878	0.9881	0.9884	0.9887	0.9890
2.3	0.9893	0.9896	0.9898	0.9901	0.9904	0.9906	0.9909	0.9911	0.9913	0.9916
2.4	0.9918	0.9920	0.9922	0.9924	0.9927	0.9929	0.9931	0.9932	0.9934	0.9936
2.5	0.9938	0.9940	0.9941	0.9943	0.9945	0.9946	0.9948	0.9949	0.9951	0.9952
2.6	0.9953	0.9955	0.9956	0.9957	0.9959	0.9960	0.9961	0.9962	0.9963	0.9964
2.7	0.9965	0.9966	0.9967	0.9968	0.9969	0.9970	0.9971	0.9972	0.9973	0.9974
2.8	0.9974	0.9975	0.9976	0.9977	0.9977	0.9978	0.9979	0.9980	0.9980	0.9981
2.9	0.9981	0.9982	0.9982	0.9983	0.9984	0.9984	0.9985	0.9985	0.9986	0.9986
z	3.00	3.10	3.20	3.30	3.40	3.50	3.60	3.70	3.80	3.90
P	0.9986	0.9990	0.9993	0.9995	0.9997	0.9998	0.9998	0.9999	0.9999	1.0000

To better present the method, let us consider the project illustrated by Table 13.2. The AON network representation below uses the expected durations of individual activities on to compute the total expected duration of the project. The total length of the critical path T_e equals 11, through activities C-E-G (Fig. 13.3).

The resulting total variance of the critical path is:

$$S^2 = V[C] + V[E] + V[G] = 0.25 + 0.25 + 0.1111 = 0.611$$

and the total standard deviation is computed as:

$$S = \sqrt{0.61111} = 0.7817$$

Table 13.2 Precedence table with expected durations and deviations for a sample project

Activity	Predecessor	a	m	b	d	v
A	–	1	2	4	2,17	0,25
B	–	5	6	7	6,00	0,11
C	–	2	4	5	3,83	0,25
D	A	1	3	4	2,83	0,25
E	C	4	5	7	5,17	0,25
F	A	3	4	5	4,00	0,11
G	B,D,E	1	2	3	2,00	0,11

Now, the probability that the project will be finished in D days or less can be computed as the probability associated with the value z in the normal distribution chart:

$$P_n(z) = (D - T_e)/S \tag{13.4}$$

where T_e equals the critical duration and S is the calculated total standard deviation.

In the sample above, the likelihood of ending the project before time 10 is 10%; in fact, it can be calculated as follows:

$$P\left(z \le \frac{10 - 11}{0.7817}\right) = P(z \le -1.2793) = 1 - P(z \le 1.2793) = 1 - 0.8997$$
$$= 0.1003 = 10\%$$

Similarly, the probability of completing the project before the expected time 13 is:

$$P\left(z \le \frac{13 - 11}{0.7817}\right) = P(z \le 2.5585) = 99.48\%$$

corresponding to the level of accuracy in determining the expected duration of individual activities.

Also, the chance associated with the expected total duration is 50% because:

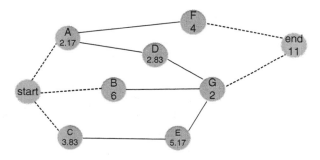

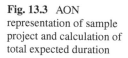

Fig. 13.3 AON representation of sample project and calculation of total expected duration

$$P\left(z \leq \frac{11 - T_e}{S}\right) = P\left(z \leq \frac{11 - 11}{0.7817}\right) = P(z \leq 0) = 50\%$$

In fact note that as D approaches T_e, z gets smaller approaching 0, corresponding to 50% probability of occurring.

This example shows the significance of accounting for uncertainty in a project. The lesson learnt here is that if one wants a reasonable chance of meeting the project deadline there must be some float in the project schedule or there must be some confidence in the probability of meeting a deadline.

Thus, PERT becomes a useful tool to determine the probability that the critical path ends within certain dates or the duration consistent with a high likelihood.

For example, the probability of the case critical path to be finished between 9 and 15 is below:

$$P(9 \leq T \leq 11.5) = P(T \leq 11.5) - P(T \leq 9)$$

$$= P\left(z \leq \frac{11.5 - 11}{0.7817}\right) - P\left(z \leq \frac{9 - 11}{0.7817}\right)$$

$$= P(z \leq 0.6396) - P(z \leq -2.5585)$$

$$= P(z \leq 0.6396) - [1 - P(z \leq 2.5585)]$$

$$= 0.7389 - [1 - 0.9948] = 0.7389 - 0.0052 = 73.37\%$$

Similarly, the deadline consistent with 95% of chance equals 12.28 time units and can be computed backwards as:

$$\begin{cases} Z = \frac{(D - T_e)}{S} \\ Z(.95) = 1.645 \rightarrow D = \sqrt{V} * Z + T_e = 0.7817 * 1.645 + 11 = 12.28 \\ S = 0.7817 \end{cases}$$

There are some disadvantages of using PERT to analyze project and activity uncertainty. Managers are required to make very accurate estimates (at one-in-a-hundred level) of individual activities durations in order to produce reasonable and reliable results from the Beta distributions. Some managers do not have the confidence to say that the optimistic and pessimistic duration guesses are exactly 99% accurate. They might be, say, only 90 or 95% confident about it.

In such cases, the degree of approximation is fostered and underestimation is avoided through a correct calculation of the standard deviation. If accuracy of duration estimates is at the 95% level:

$$Z(0.95) = 1.65$$

so that b–a = 2 (1.65) = 3.3 S; S = range/3.3
In case of 90% accuracy:

$$Z(0.90) = 1.3$$

b–a = 2 (1.3) = 2.6 S; S = range/2.6

And so on for more approximate estimates of optimistic and pessimistic durations.

Furthermore, the PERT can lead to unrealistic determination of expected completion times if the sole critical path is taken into account. A better calculation of probability associated with the total expected duration should consider that any time two or more paths merge, the probability of both paths being on time is the product of probabilities for the individual paths. Particularly, the problems of optimistic estimation incur when sub-critical path have low float and high variance. The principle goes under the name of "merge node bias".

Thus, the correct estimation of the expected duration of our case network is the following (Fig. 13.4).

Assuming that the critical path Start-C-E-G-End has 50% probability associated with its expected duration 11 and that the sub-critical path Start-B-G-End has variance equal to (from values in Table 13.2):

$$V = 1 + 0.11 = 1.11; S = 1.05$$

and probability of ending before 11 equal to $P = 98.68\%$, where $z = (11-8.67)/1.05 = 2.22$, and assuming independent paths in the network, it is possible to compute the probability of both paths to be completed before 11 as:

$$P (B - G) * P (C - E - G) = 98.68\% * 50\% = 49.34\% < 50\% \text{ critical path only}$$

Briefly, it is misleading to consider only variance from single predecessor for each node on the critical path because the early start of a node depends on maximum of finish (or start) times of predecessors, including the non-critical. The effect of the merge node bias is even stronger if the node has more predecessors, predecessors with almost equal timing and if there is dependency among predecessors. The consequence is an unrealistic optimism with respect to expected completion times, but especially to variance (Moder et al. 1983).

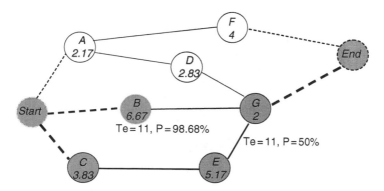

Fig. 13.4 AON representation of sample project and calculation of total expected duration

Thus, regardless the usefulness, PERT evaluation has limitations associated with statistical assumptions and inherent disadvantages.

The validity of Beta distribution for activity durations and the central limit theorem for the project duration is questionable because PERT treats each event autonomously, while most often activity durations are not independent. Also, the critical path is considered as a sum of random variables, which is technically inaccurate. This does not mean the project schedule will become infeasible, but it does generally lead to over-optimism and underestimation of project duration: slight estimation errors of the optimistic value, a, or the pessimistic value, b, can skew the entire distribution.

This makes for repeated, long calculations to be completed every time a new duration value is chosen.

For this reason, some project managers have moved to even more advanced algorithms for measuring uncertainty. Both Monte Carlo, Graphical Evaluation and Review Technique and other simulations are such methods.

13.3 Simulations

Discrete-event simulations, such as Monte Carlo, are valuable tools to better calculate project durations associated with probability. Moreover, transaction-based modeling is available to deal with uncertainty and risk when projects are not only uncertain in the duration, but also highly dependent on other variable conditions and constraints. GERT and Q-GERT use activity-cycle diagrams to perform simulations of the activity flow in a network with looping and branching.

13.3.1 Monte Carlo Simulations

To overcome several PERT disadvantages, a Monte Carlo simulation can be run to determine the expected project duration and cost to completion.

The purpose of a Monte Carlo simulation (see also Chap. 12) is to replace analytic solving with raw computing power in order to avoid the need to simplify to get analytic solution and to assume functional form of activity distributions. Simulations are aimed at solving the merge bias problem contained in PERT analysis and allow determining the probability of the project to be finished before a deadline. This probability is assumed as the proportion of runs in which the project ends before the deadline and the total number of simulations (Hendrickson 2008).

This requires hundreds to thousands of simulations, which is typically not a problem on today's computers. With increased computer processing speed, Monte Carlo simulations take very little time to run.

A simulation first starts with a network of activities (Fig. 13.5). Each activity has a corresponding value for optimistic, most-likely, and pessimistic duration as well as the expected duration and standard deviation (Table 13.3).

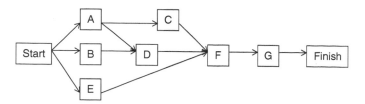

Fig. 13.5 Sample of AON network

Table 13.3 Sample of AON network

Activity	Optimistic duration, a	Most likely duration, m	Pessimistic duration, b	Expected duration, d	Standard deviation, S
A	2	5	8	5	1
B	1	3	5	3	0.67
C	7	8	9	8	0.33
D	4	7	10	7	1
E	6	7	8	7	0.33
F	2	4	6	4	0.67
G	4	5	6	5	0.33

With these values into the computer, the simulation is ready to go. During the simulation the computer completes a preset number of iterations, also called trials or realizations. Each run, the value for the activity duration is chosen randomly from a Beta distribution and then the critical path expected duration is computed with standard CPM. These results are recorded in Table 13.4.

Once a number of runs have been performed, it is possible to draw a distribution chart and calculate the probability associated with a given deadline as the ratio between the number of trials the project finished before the deadline and the total number of runs (e.g. from Table 13.4, P(<20) = 3/10 = 30%).

Also, each node, which represents one particular activity, is referenced by a criticality index. This index is defined as the proportion of realizations in which that

Table 13.4 Monte Carlo simulations of sample project – 10 realizations

Run #	A	B	C	D	E	F	G	Critical path	Completion time
1	6.3	2.2	8.8	6.6	7.6	5.7	4.6	A-C-F-G	25.4
2	2.1	1.8	7.4	8.0	6.6	2.7	4.6	A-D-F-G	17.4
3	7.8	4.9	8.8	7.0	6.7	5.0	4.9	A-C-F-G	26.5
4	5.3	2.3	8.9	9.5	6.2	4.8	5.4	A-D-F-G	25.0
5	4.5	2.6	7.6	7.2	7.2	5.3	5.6	A-C-F-G	23.0
6	7.1	1.0	7.2	5.8	6.1	2.8	5.2	A-C-F-G	22.3
7	5.2	4.7	8.9	6.6	7.3	4.6	5.5	A-C-F-G	24.2
8	6.2	4.4	8.9	4.0	6.7	3.0	4.0	A-C-F-G	22.1
9	2.7	1.1	7.4	5.9	7.9	2.9	5.9	A-C-F-G	18.9
10	4.0	3.6	8.3	4.3	7.1	3.1	4.3	A-C-F-G	19.7

specific activity was in the critical path. This is very helpful in optimizing the schedule prioritization, as well as project monitoring and controlling. An activity with a high criticality index value will be closely monitored so to ensure the project does not slip behind schedule as a result of that activity.

In the sample project, the criticality index of activity A is the proportion of times that node A is on the critical path. In this simulation of only ten realizations, it is on the critical path every time. Therefore, the criticality index of activity A is 100%. Activities B and E never make it into the critical path so they have criticality indices of 0%. Activity C has eight of ten appearances in the critical path for a criticality index of 80% and Activity D has a criticality index of 20%.

From this limited number of realizations, one can deduce which activities are most likely to be along the critical path: A, F, G and then either C or D. But to get more precise results, one should consider doing more than ten trials. Errors of estimation are present when only a few trials are considered. To reduce the error, more trials must be simulated. The empirical rule is that the mean project duration will be within 0.5–1.5% of the exact value for 400 realizations and within 0.3–1% for 1,000 realizations. The standard deviation of the project is not quite as accurate, but 400 realizations will keep it within 7% of its actual value. A simulation of 1,000 trials will limit the error of the standard deviation to 4.4% or less.

Most users of the Monte Carlo simulation agree that 1,000 realizations are adequate for accurate results. To exemplify the results from a 50-realization test, Fig. 13.5 is a histogram that shows the relative frequency with which each of the durations occurred. One can see how the normal distribution (symmetric Beta distribution) takes shape as more and more realizations are recorded. The average duration of the critical path is 22.6 days with a standard deviation of 2.95 days (Fig. 13.6).

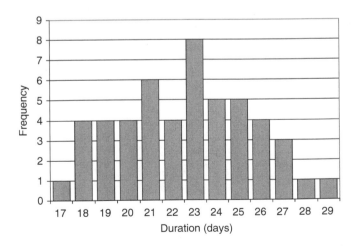

Fig. 13.6 Monte Carlo simulation of 50 realizations

In summary, it is imperative that the uncertainty of project activities be assessed both for expected duration and variance. For any large-sized construction project, uncertainty analysis is a large part of the controlling and monitoring that a manager must do. A project that overruns its time schedule will cost the owner or the contractor lots of money in liquidated damages. The reputation of the owner or contractor could also be compromised for future business dealings (Pritsker and Sigal 1983).

A simple, indifferent approach of ignoring uncertainty or adding a buffer may not be sufficient and a good project manager will prepare an appropriate probabilistic schedule to manage a large-sized and complex construction project.

13.3.2 GERT

Graphical Evaluation and Review Technique is a network modeling for high-complex projects where CPM or PERT are unsatisfactory to provide an accurate scheduling under uncertain conditions, ongoing decision making and process options. Essentially, it is an activity-on-arrow network, which specifies the sequencing of project tasks, with added decision-making information in the nodes.

In fact, the nodes of a GERT, and more extensively of a Queue-GERT, network may represent events, storage points (queues), routing decisions and resource allocation decisions. By building a network of nodes connected by arrows, complex situations can be modeled: the activities of the project network can be performed under varied conditions associated with the nodes. In other terms, a GERT is a network model very close to a PERT with added decision trees and process flow graphs: activities may unfold through different branches according to the result in the decision nodes, and may also be repeated in loops to simulate, for example, the process of redesign and rework (Taylor and Moore 1980).

GERT not only assumes probabilistic durations but also that branching and looping from a node are also probabilistic.

Activities are specified by a function type and parameter identifier. An activity with deterministic duration is identified by a constant function and its specified duration (i.e. CO, 10 days). A probabilistic activity is described by a distribution function and the identifier of a set of values. For example, a function type may be a Beta distribution and the identifier refer to a first set of frequency values available in the system library (i.e. BE, set 1). This would specify that the activity duration should be a random sample from the beta distribution the modeler has defined.

At the end of each activity is a node, referred to as a rooting point. The flow of activities depends on the rooting specifications prescribed for the node.

Basically, four types of rooting can be specified at a node:

- deterministic routing allows for duplicating as well as merging paths of activities, similar to CPM nodes;
- probabilistic rooting entails the selection of the successor from a set of activities departing from the node; the selection is based on the relative frequency with which the activity is statistically performed;

- conditional rooting selects the activity emanating from the node based on a pre-scribed condition, such as AND, OR, or more complex situations (i.e. take the first activity that satisfies the condition or take-all the activities that satisfy the same condition).

Also, there are four main types of node to model the decision process, milestones, and events associated with the flow of activities:

- basic nodes are tools that allow for keeping track of the flow, such as marking times or make statistics for the simulation,
- queue nodes represent buffer areas where the process is halted until a resource is available to the node via an incoming activity,
- selector nodes allow for modeling rooting specifications and priority rules.

As a general consequence, GERT involves more extensive information and compu-tational requirements.

The method requires to, first, define the network as a PERT graph using proba-bilistic distribution of durations; second, add chance/decision nodes and other logic relationships to the base network; then, solve the network to yield the probability of each node being realized, the times when nodes are possibly realized and the time between all nodes; and finally, analyze results and make inferences on the simulation (Pena-Mora and Li 2001).

Modeling a project with GERT is certainly a time-consuming activity that is recommended for complex situations that involve high levels of uncertainty. In par-ticular, it is of practical use as a decision-making and decision-analysis tool and as a project planner in the face of risk and uncertainty

References and Additional Resources About Probabilistic Scheduling

Goldratt EM (1997) Critical chain. North River Press, Great Barrington, MA

Hendrickson C (2008) Project management for construction, 2nd edn. Carnegie Mellon University, Pittsburgh, PA. http://pmbook.ce.cmu.edu/11_advanced_scheduling_techniques.html

Krishnan R (1998) PDM: a knowledge-based tool for model construction. Proceedings of the 22nd annual Hawaii conference on construction, Kailua-Kona, HI

Moder JJ, Phillips CR, Davis EW (1983) Project management with CPM, PERT, and precedence diagramming. Van Nostrand Reinhold, New York, NY

Pena-Mora F, Li M (2001) Dynamic planning and control methodology for design/build fast-track construction projects. J Constr Eng Manage 127(1):1–17

Pritsker AAB, Sigal CE (1983) Management decision making a network simulation approach. Prentice Hall, Englewood Cliffs, NJ

Schtub A, Bard J, Globerson S (1994) Project management: engineering technology, and imple-mentation. Prentice Hall, Englewood Cliffs, NJ

Taylor BW, Moore LJ (1980) R&D project planning with Q-GERT network modeling and simulation. Manage Sci 26(1):44–59

Chapter 14
Risk Management

14.1 Risk Identification

The British Standard Institute (1991) defines risk as "a combination of the probability of occurrence of a defined hazard and the magnitude of the consequences of the occurrence", or as a combination of likelihood of occurrence of a certain problem with the corresponding value, i.e. impact, of the damage caused.

The definition implies that there are two basic components to risk: probability of an event occurring and the negative impact due to the occurrence of the event. These two basic components are essentially independent, but are used in unison to categorize risk.

For example, take a simple risk categorization that classifies risk as either acceptable or unacceptable on a construction site. An acceptable risk for a task could have a probability of occurrence of 1/1,000,000 and a negative impact resulting in the fatality of a worker. An unacceptable risk for a task could have a probability of occurrence of 1/100 and a negative impact of the loss of a finger for the worker. Categorization of risk is a much more developed and varying process across industries.

Construction risks may be classified into three broad categories: financial, schedule, and design (Macomber, 1989). The most basic financial risk is when there are project cost overruns that impact on the financial strength of the contractor, owner, developer, or whoever bears this risk. Financial risks are not necessarily due to internal errors, but for example could result from the insolvency of a subcontractor. Schedule risks stem from the project not being completed on time, which in turn has drastic financial impacts. The final construction risk generates from the design of the building. This occurs when the completed building does not meet the needs of the owner and occupants. Changing needs of the owner over time or poor communication between the design staff and the owner creates this risk.

Financial, schedule and design risks are of prime concern for the project team, who must be aware that the sources of risks are complex and varied.

Risks may be originated by internal factors, namely in the main areas of contract, people, and material/technology (ironically, as the structure of this textbook!), or by external ones, such as natural conditions (weather, etc.), economic situation and political context.

A. De Marco, *Project Management for Facility Constructions*,
DOI 10.1007/978-3-642-17092-8_14, © Springer-Verlag Berlin Heidelberg 2011

As a principle, internal sources of risk are project-originated so that the project team may actively control them; external sources are not under the direct control of the project team.

Take for example inclement weather: the project team must be aware of both the financial risk as well as the schedule risk this presents. Once these two categorizations are understood the team can focus on identifying these risks.

In identifying construction risk, three elementary inputs have to be considered.

The first and most important is the project itself. Considerations within the project include the type of building or structure, the project objectives, project requirements, constraints and limitations, and surrounding site conditions among others.

The second input is the management system being utilized. These include methods, tools and practices.

Finally is the context of the stakeholders involved: developers, landlords, construction managers, project managers, prime contractors, subcontractors, material providers, etc. The background information for each company should be considered such as experience from past projects, historical information, and resources they are allocating to this project. Though this is not an exhaustive list of necessary inputs in comprehending construction risk, gathering as much information as possible is the first step in managing risk.

Several techniques may be used to investigate and identify possible causes of risk in a project.

Interviewing experts and project managers with specific experience in similar projects is a valuable basic step. Also, standard checklists may be available to start identification from a panel of frequent risky events based on corporate past experience.

Another useful way in identifying risk is a "what-if" analysis. This step involves asking a series of, "what would happen if. . ." questions. The goal of this analysis is to consider all potentially risky situations in the project and to be able to understand the consequences either qualitatively or quantitatively. Various styles of cause and effect diagrams can be used to understand the consequences of occurrences.

Event Tree Analysis (ETA) and Fault Tree Analysis (FTA) are two examples of available options. ETA uses a bottom up approach: causes are analyzed and their potential effects are determined. FTA uses a top down approach: an undesirable event is considered and all of the possible ways it can happen are determined. Each methodology is limited in its application, and thus several are often used for an individual project. Other methodologies are available, including failure modes effects, and criticality analysis (Chapman and Ward, 2003).

ETA begins with the identification of a potential risky event that will have a negative outcome on the project. Further events that may occur as a result of the first risky event are also determined. An event tree is finally drawn and the probabilities of each path are determined.

FTA proceeds by determining how the top level event can be caused by individual or combined lower level causes. The tree depends on the use of various symbols such as logic gates (AND, OR, etc.), input events, description of states and so on.

Fig. 14.1 Example of fault tree analysis

The application of FTA in investigating the causes of an unattended schedule is seen in Fig. 14.1.

14.2 Risk Breakdown Structure

A Risk Breakdown Structure (RBS) allows the project team to classify risky events in a hierarchical system, similar to a WBS.

The RBS should be used in conjunction with the above "what if" analysis methods to determine potential sources of risk. The elementary causes of risk stem from the bottom of the tree. According to the breakdown structure, the set of risky events (causes) is initially split in risk types; each type is in turn subdivided into classes, which are further decomposed into groups, sub-groups, and so on down to the basic elementary risky event.

The typical RBS for a construction project is given in Fig. 14.2.

14.3 Risk Quantification

Beyond identifying risks, there is a need to quantify the impact of unforeseen effects. Simple ways of quantification will facilitate the creation of contingency plans, as well as adequate contingency budgets in order to confront risky events when they occur.

To quantify the consequences of a specified risk on an activity it is required to draw a matrix relationship between all elementary risk sources and the WBS activities (Hillson et al., 2006), as shown in Fig. 14.3.

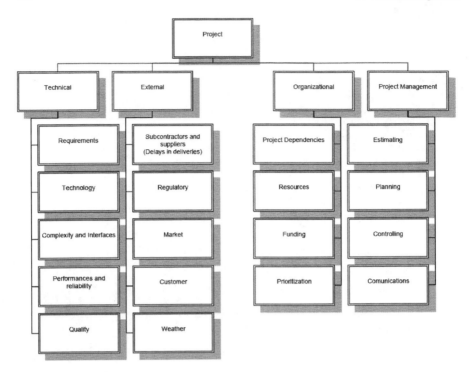

Fig. 14.2 Example of a risk breakdown structure for a construction project

Fig. 14.3 Matrix between
tasks and risks

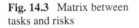

		WBS				
		A1	A2	A3	...	AJ
	E1		R_{12}			
	E2	R_{21}		R_{23}		
	E3		R_{32}	R_{23}		R_{3J}
	...					
	EK	R_{k1}				R_{kJ}

At the cross point, the monetary effect may be estimated as the risk exposure (R) of activity j being subject to the risky event k.

$$R = p * I$$

where p is the probability that the negative event could occur and I is the economic impact of this event, usually measured as a financial loss.

R is a useful indicator in the ranking of risks to be addressed. Assume for example the two tasks mentioned above where task 1 results in the fatality of a worker (p = 1/1,000,000) and task 2 results in the loss of a limb (p = 1/100). The financial impact due to the fatality of a worker could be approximated at $10 million and the impact due to the loss of a limb at $100,000. Using the formula given above:

$$R \text{ (task 1)} = 1/1,000,000 * \$ 10,000,000 = \$ 10$$

$$R \text{ (task 2)} = 1/100 * \$ 100,000 = \$ 1,000$$

Though this might seem like an oversimplification of the problem, analysis of this type is what is necessary in order to focus the project team in a manner such that they minimize risk for the project as a whole. In this case, the project team would be willing to accept risk 1 more than 2.

There are three approaches to quantifying risk exposure elements: qualitative, semi-qualitative, and quantitative.

Qualitative quantification relies on the use of a range of "word" values for the risk presented, using various levels for the probability of that risk and its corresponding impact. For example, a simple qualitative risk evaluation may use the following values for p and I:

- p: very high, high, medium, low, very low
- I: catastrophic, critical, medium, marginal, negligible.

A semi-qualitative approach is very similar to a qualitative approach, but the descriptive levels are classified numerically. The chart below is an example of how to assign numerical values to the "word" values from Table 14.1.

The assigning of numerical values allows for easier analysis of the individual risks as well as for classification of risks. The use of a fully quantitative approach facilitates risk ranking most effectively.

Quantitative quantification of risk relics purely on the use of numbers. The probability of an occurrence is usually given as a simple percentage, unless more accurate data is available in order to create a probability distribution for the occurrence of the event. The impact of the event is measured in regards to various project parameters such as cost, time, or performance level. For example, the risk associated with a construction task could be a delay of four working days – in this case the number of days would be used as a time parameter to calculate extra cost.

Table 14.1 Semi-quantitative assessment of risk exposure

Probability				
Very high	High	Medium	Low	Very low
5	4	3	2	1
Impact				
Catastrophic	Critical	Medium	Negligible	Insignificant
5	4	3	2	1

ACTIVITY		O.B.S.	PERIOD OF PERFORMANCE		RISK DESCRIPTION	RISK EXPOSURE			COUNTER MEASURES		
COD	NAME		START	END		I	P	$	DESCRIPTION	obs	$

Tot. ☐   Tot. ☐

Fig. 14.4 Risk assessment report and estimation of the contingency budget

Quantitative assessment of risk exposure requires a large number of historical data to estimate probabilities of occurrence of risks and associated economic impacts if risks happen.

The final output of risk quantification is a Risk Assessment Report (Fig. 14.4).

Below is a template of a risk assessment report: for each activity from the WBS, one or more risk events generate a monetary risk exposure and a cost of counter measures.

This allows for estimating a proper contingency budget to be used as a cost buffer if risks will occur during the project execution. The contingency budget equals the summation of the cost of risk for all project activities, where the cost of risk is the minimum between the risk exposure (R) and the cost of the preventive action (A) that is needed to face the risk: it is clear that it is preferable to bear the risk if the action needed to minimize or cancel it is more expensive than the risk exposure. The formula is below:

$$\text{contingency budget} = \sum_k \sum_i [\min (R_{ik}; A_{ik})]$$

In the process of quantifying risk exposure and the cost of counter measures it may be opportune to take into account co-relations between effects of the same risk on more than one activity: for example, if a risky event likely to provide negative consequences on several activities occurs on an early activity, it may not have effect on a late one. In this case, the risk exposure or the cost of the preventive action has to be considered only once. Also, the preventive action may only contribute in part to reducing the risk, so that it may be convenient to consider as a contingency both the cost of the countermeasure and the remaining monetary risk exposure. Of course, these are refinements that try to better estimate uncertainty.

In conclusion, the process of identifying and quantifying contingency budgets based on a detailed breakdown is certainly an approximation of the problem, but it still provides, on the one hand, a better understanding of the project risks than simply calculating the contingency as a percentage of the total budget and, on the other hand, a more practical approach than sophisticated model simulations. Simply estimating a contingency as a percent cost of the total budget would fail in underestimating or overestimating the amount of money and time to use as buffers for pricing and scheduling the contract, while making more accurate models using probabilistic simulations or GERT approaches may be appropriate for complex large-scaled projects, and therefore perceived as useless by most project managers involved in small and medium-sized projects.

14.4 Risk Control

Only once the risk on a project is understood, identified, and quantified the project team can take appropriate steps to manage risk.

Basically, responses to be used towards risk may be drawn in four directions: avoid, transfer, mitigate, and accept risk.

All of these involve a good understanding of the contract and impact on the risk attitude of the parties to shoulder the risk (see Chaps. 3 and 12). The diagram in Fig. 14.5 will help determine which response to pursue depending on the probability and impact of the risk.

It is abundantly clear that a low-likelihood/low-impact risk is eager to be accepted by both parties in a contract. Different risk attitudes may exist for cases when either

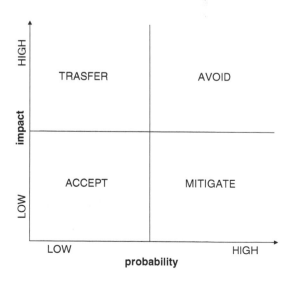

Fig. 14.5 Main types of risk control strategies

the probability or the impact (or both) have high values. Therefore, the guide-lines provided below for each type of control action have to take into account all the elements of the contract negotiation with regard to the owner/contractor attitudes to risk:

- Avoid – Avoiding risk is the most simplistic and often the most practical method by which risk can be minimized. Avoiding risk will possibly include changing project objectives and possibly considering alternative solutions. For example, using a new and unfamiliar construction technique poses tremendous risk. Reverting to the traditional method will avoid this risk.
- Transfer – Transferring risk to other stakeholders is a second basic option. The use of insurance policies will transfer risk onto the insurer. The obvious downside is the risk premium, which represents the cost of the counter measure, as presented in the previous paragraph. Types of insurance common in the construction industry include general builder's risk insurance as well as general liability insurance. Non-insurance transfers can be completed through the hiring of sub-contractors as well as through contract clauses.
- Mitigate – Mitigation involves a range of activities designed to reduce project risk. These activities include scheduling risky tasks out of the project critical path, allocating resources in order to minimize negative impacts, as well as holding frequent update meetings on important project aspects among others.
- Accept – The least desirable response is to accept full risk. Even though the risk is "accepted" there are still options available in order to minimize the impact from this risk. Monitoring plans devoted to those risky activities should be created. These plans should consider recovery and determined resources necessary to mitigate the impacts. For each risk an accurate probability of the risk occurring as well as its impact (financial or schedule) must be determined. Counter measures must be clearly defined– actions to be taken, responsible persons for initiating these actions, and the residual effects even after the actions are taken. In addition, contingency funds or materials, depending on the task, should be allocated to provide for a proper response. Finally, accepting risk may be an interesting option if a proper risk premium is paid to the party who shoulders the risk (see Chap. 3).

References and Additional Resources About Risk Management

British Standard Institute (1991) Quality vocabulary. Availability, reliability and maintainability terms. Guide to concepts and related definitions. No. 4778, British Standard Institute, London
Chapman CB, Ward SC (2003) Project risk management: process, techniques and insights, 2nd edn. Wiley, Chichester
Haimes YY (2004) Risk modeling, management, and assessment. Wiley, Hoboken, NJ
Hillson D, Grimaldi S, Rafele C (2006) Managing project risks using a cross risk breakdown matrix. Risk Manag 8:61–76

Macomber JD (1989) You can manage construction risks. Harvard business review. Harvard University, Cambridge, MA, March–April 1898

Project Management Institute (2008) A guide to the project management body of knowledge, 4th edn. Project Management Institute, Newtown Square, PA

Ugur M (2005) Risk, uncertainty and probabilistic decision making in an increasingly volatile world. Handbook of Business Strategy 6(1):19–24

Abbreviations and Acronyms

A/E	Architect/Engineer
AC	Actual completion date
ACWP	Actual cost of work performed
AEC	Architecture engineering and construction
AOA	Activities on arrows
AON	Activities on nodes
BAC	Budget at completion
BC	Budgeted completion date
BCWP	Budgeted cost of work performed
BCWS	Budgeted cost of work scheduled
BIM	Building information model
BOT	Build operate and transfer
CAD	Computer aided design
CBS	Cost breakdown structure
CI	Cost index
CM	Construction manager
CP	Critical path
CPM	Critical path method
CV	Cost variance
D/B	Design-build
DBB	Design build bid
DCF	Discounted cash flow
DSCR	Debt service coverage ratio
EAC	Estimate at completion
EMV	Expected monetary value
EPC	Engineering procurement and construction
ERP	Enterprise resource planning
ETA	Event tree analysis
EUV	Expected utility value
EV	Earned value
EVA	Earned value analysis
FTA	Fault tree analysis

A. De Marco, *Project Management for Facility Constructions*,
DOI 10.1007/978-3-642-17092-8, © Springer-Verlag Berlin Heidelberg 2011

FV	Future value
GC	General contractor
GDP	Gross domestic product
GERT	Graphical evalutation and review technique
GMP	Garanteed maximum price
IRR	Internal rate of return
MARR	Minimum attractive rate of return
MRP	Material requirement planning
MTO	Material take off
NPV	Net present value
O&M	Operations&maintenance
OBS	Organization breakdown structure
PDM	Precedence diagramming method
PERT	Program evaluation and review technique
PI	Project index
PM	Project manager
PMO	Project management office
POC	Project organization chart
PPP	Public private partnership
PV	Present value
RBS	Risk breakdown structure
RFC	Ready for construction
RFI	Request for information
RFP	Request for proposal
RI	Resource flow index
RV	Resource flow variance
SI	Schedule index
SPE	Special purpose entity
SPV	Special purpose vehicle
Sub	(subs) subcontractor (subcontractors)
SV	Schedule variance
SWOT	Strengths weaknesses opportunities and threats
T	Time now
TQM	Total quality management
UV	Utility value
VAC	Variance at completion
WACC	Weighted average cost of capital
WBS	Work breakdown structure
WP	Work performed
WS	Work scheduled

Index